ENVIRONMENTAL SCIENCE
AND
ENGINEERING

ENVIRONMENTAL SCIENCE
AND
ENGINEERING

T. Meenambal

Assistant Professor
Department of Civil Engineering
Government College of Technology
Coimbatore
Tamilnadu

MJP PUBLISHERS

Cataloguing-in-Publication Data

Meenambal, T. (1959 –   ).
Environmental Science and Engineering / by
T. Meenambal. –
Chennai : MJP Publishers, 2009.
    xviii, 388 p.; 21 cm.
Includes Glossary, Bibliography and Index.
    ISBN 978-81-8094-058-3 (pbk.)
    1. Ecology 2. Environment 3. Engineering, Environment
577:620 dc 22 MEE MJP 062

ISBN 978-81-8094-058-3
© Publishers, 2009
All rights reserved
Printed and bound in India

MJP PUBLISHERS
47, Nallathambi Street
Triplicane
Chennai 600 005

Publisher : J.C. Pillai
Managing Editor : C. Sajeesh Kumar
Project Editor : P. Parvath Radha
Acquisitions Editor : C. Janarthanan
Editoral Team : B. Ramalakshmi, N. Pushpa Bharathi,
Lissy John, M. Gnanasoundari,
R. Magesh
CIP Data : Prof. K. Hariharan, Librarian
RKM Vivekananda College, Chennai.

This book has been published in good faith that the work of the author is original. All efforts have been taken to make the material error-free. However, the author and publisher disclaim responsibility for any inadvertent errors.

To

My beloved Mother
with sincere respect, humbleness
and regards

PREFACE

But ask the animals, and they will teach you, or the birds of the air, and they will teach you: or speak to the earth, and it will teach you, or let the fish of the sea inform you. Which of these does not know that the hand of the Lord has done this? In His hand is the life of every creature and the breath of all mankind.

The Bible, Job 12:7-10

Natural resources are *God's gift* to humankind. It is the utmost duty of humans to preserve and conserve nature and the *Earth.* The unsustainable use of the resources by humans has caused many irretrievable losses to the earth. The interaction between humans and the natural resources has led to environmental imbalances and so we need to be judgmental in using the environmental resources for our day-to-day use. In the study of environmental science, it is important to have a historical perspective, appreciate economic and political realities and recognize the role of different social experiences and ethical backgrounds. This book covers all the information relating to the environment and its conservation. Each topic is discussed around the resources of the environment and the role of humankind in managing these resources. The major issues relating to environmental degradation are dealt with using appropriate examples and case studies.

This book is divided into four parts. Part I covers natural resources like land, water, forest, mineral, food and energy. Each chapter is are discussed in detail with all the basics involving the formation of resources and the factors affecting the resources followed by the explanation with specific case studies and examples to provide the reader with varied scope of the subject and in-depth knowledge of the facts by providing conservation management techniques for each of the resources in specific. Part II covers the ecosystem and biodiversity. It includes the structure and function of an ecosystem, the food chain and food web, the types of ecosystem and the biogeographical classification of India and biodiversity in detail. Part III discusses in detail the various types of pollution such as air, water, soil, marine, noise and thermal. The management of natural disasters, waste managtement and the role of an individual in the prevention of pollution are also discussed. Part IV discusses the social issues and the environment. Issues such as resettlement and rehabilitation, climatic change, consumerism and waste products, and environmental ethics are covered in detail. The book also includes the Environmental Protection Act and Forest Conservation Act. All the topics are discussed with appropriate case studies to understand the effects of environmental degradation and the need for conserving the natural resources to lead a harmonious life.

It is sure that this book with its systematic organization and presentation would fill up the gap of a much needed textbook in Environmental Science and Engineering for students in the under-graduate and post-graduate programmes.

I express my gratitude to all who have helped me directly or indirectly in writing this book, especially to those who inspired me to write this book.

I would like to thank the editorial team of MJP Publishers for their active cooperation and patience in bringing out this book.

T. Meenambal

CONTENTS

PART II   ECOSYSTEM AND BIODIVERSITY

PART III   ENVIRONMENTAL POLLUTION

13. Air Pollution   197

PART IV   SOCIAL ISSUES AND THE ENVIRONMENT

Man lives in nature and depends on the resources of nature. The progress of mankind depends upon the utilization of different natural resources such as soil, water, coal, electricity, oil, gas, nuclear energy, etc. The utilization of these resources is very important for the development of a nation. They have changed the living standards of man. India has the world's largest resource of coal and the third and fourth largest resource of manganese and iron.

The earth is facing an ecological crisis and her natural resources are degraded day by day due to over-exploitation. India is no exception. Food, shelter, and clothing are the primary requirements of man. Early human society has used the natural resources, relatively in much less quantity to cover only its needs.

DEFINITION, SCOPE AND IMPORTANCE

The term **resource** means a source of supply or support that is generally held in reserve. It is greatly needed by the humans for their survival and comforts. The components of lithosphere, hydrosphere, and atmosphere constitute the natural resources, e.g. energy, air, water, soil, minerals, plant, and animals.

In order to achieve a proper utilization of natural resources, the ingenuity of human beings plays a vital role. With the technological innovations in extraction and processing of natural resources to furnish materials, human beings have learnt to locate and extract these resources. The furnished materials from these natural resources constitute the base of our material wealth.

The economic exploitation of non-renewable resources during the last two centuries has, in general, been one of the factors in the decreasing costs, increasing reserves, and rising energy and work cost requirements. Some scientists think that most of our mineral and energy resources are finite, and yet the world is not likely to run out of such resources: they will merely become more and more expensive.

Of late, we have come to realize the fact that we have to keep pace with the exploding population in the matter of increasing productivity in all fronts and that we should try judiciously to utilize the available resources to the best advantage so as to prevent the destruction of these available resources. It must also be emphasized that any imbalance in the exploitation of natural resources may result in an irreversible damage to the quality of the life-supporting environment.

TYPES OF NATURAL RESOURCES

Natural resources are classified in two ways.

Classification based on Chemical Composition

Based on their chemical composition, natural resources are classified into three types. They are

1. Inorganic resources, e.g. air, water and minerals
2. Organic resources, e.g. plants, animals, microorganisms, and fossil fuels
3. Mixture of inorganic and organic resources, e.g. soil

Classification based on Availability and Distribution

Based on their availability and abundance, they are classified into two types. They are:

Inexhaustible resources   They are the resources that are not likely to be exhausted by human use, e.g. air, clay, sand, tidal energy, etc. Although air cannot be exhausted, it can be degraded if its pollution is not checked.

Exhaustible resources   They are the resources that are likely to be exhausted by human use. They are further grouped into two types, namely, renewable and non-renewable resources (Figure 1.1).

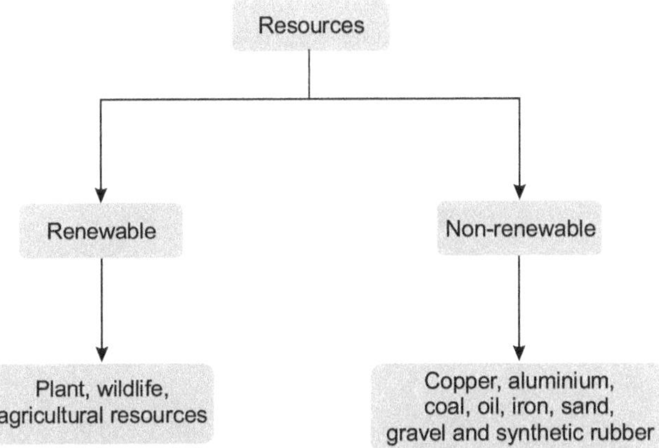

Figure 1.1   Renewable and non-renewable resources

Renewable resources   They have an inherent capacity to reappear or replenish themselves by quick recycling, reproduction, and replacement within a reasonable time. Soil and living organisms are the main renewable resources. They include plants and animals and are often referred to as bioresources or living resources, e.g. plant resources, wildlife resources, fishery resources, agricultural resources and forest resources, medicinal plants, etc. They are crucial to an enduring human civilization.

Non-renewable resources   They lack the ability for recycling and replacement. The substances with a very long recycling time are also regarded as non-renewable resources, e.g. fossil fuels like coal, petroleum and natural gas, and minerals. They consist of geochemical concentrations of naturally occurring elements and compounds that may be exploited profitably and are grouped under three categories.

    i.   Non-metallic minerals, e.g. sand and gravel, stone, cement, and clay

   ii.   Metals, e.g. iron, steel, aluminum, copper, lead, zinc, and others

  iii.   Polymers, e.g. plastics and resins, synthetic rubbers and non-celluloid fibres

A basic knowledge of the following natural resources is necessary so that it can be properly utilized and conserved for future.

1. Forest resources
2. Water resources
3. Mineral resources
4. Food resources
5. Energy resources
6. Land resources

These are discussed in detail in the following chapters.

The forest is a complex ecosystem consisting mainly of trees that buffer the earth and support a myriad of life forms. The trees create a special environment that affects the kinds of animals and plants that can exist in the forest. They clean the air, cool it on hot days, conserve heat at nights, and act as excellent sound absorbers. Thus, trees are a very important component in the environment.

FORESTS AND THE ENVIRONMENT

Forests are important natural resources. It is the most important natural habitat for wildlife, where herbivores find their shelter and carnivores their prey. Animals, insects and other organisms live in various parts of the plant, e.g. birds build their nests on the branches of trees. Besides this, they play an important role from commercial point of view. Forest-based cottage industries, such as bee keeping, bamboo mat and basket-making, provide a small-scale business to the tribal people. It also serves for the commercial and recreational purposes that can be utilized by the farmers. Sal is the most important source for timber industries. It also provides raw materials for pulp and plywood industry.

Green plants of the forest are the food-producing organisms and are termed the primary producers of the "food chain". These foods are stored up in the form of fruits, nuts, seeds, nectar, and wood. Therefore, the forest serves as an energy reservoir by trapping energy from sunlight and storing it in the form of a biochemical product.

Forests play an important role in maintaining the atmospheric balance by consuming the CO_2 present in air and by producing the O_2 that is very essential for all the living beings. (An acre of forest absorbs four tons of carbonic acid gas and recycles eight tons of oxygen into environment.) Therefore, the removal of plants and trees would disturb the composition of natural air. Transpiration, i.e., giving off water vapour through the stomata, affects the relative humidity and precipitation in a place.

Forests prevent soil erosion by binding the soil with a network of roots (Figure 2.1). They provide a protective cover (**canopy**) that reduces the impact of rain on the soil, thereby reducing soil erosion. The tree roots stabilize the soil and aid a proper water flow.

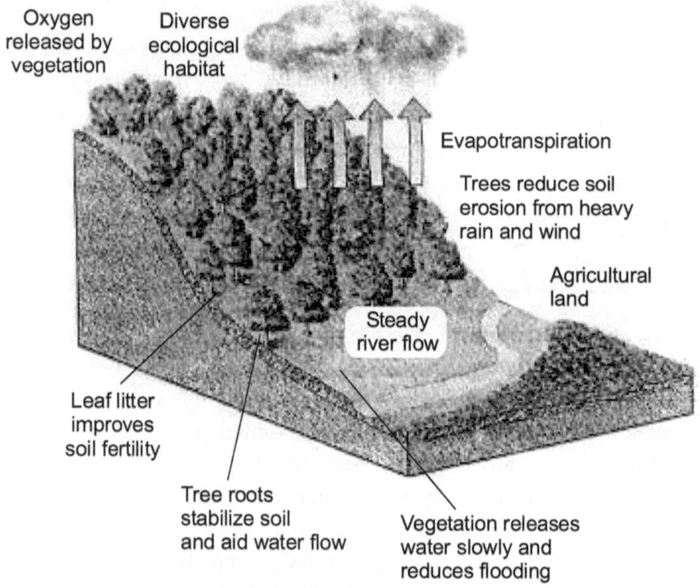

Figure 2.1   Role of forests

Dead plants or leaf litters decompose to form humus (the organic matter that holds the water and provides the nutrients to the soil) thereby enhancing the soil fertility. In the forests, the thick layer of humus acts like a big sponge and soaks the rainwater preventing runoff, thereby preventing floods. It also prevents the quick evaporation of water thereby ensuring a perennial supply of water to streams, springs, and wells.

Forests are the main source of wood and bamboos. Wood is used as an energy source for cooking purposes and for keeping warm. It is also used for making **timbers** such as furniture, tool-handles, railway sleepers, matches, boats, etc. **Bamboos** are used for making baskets, ropes, rafts, cots, etc. Both bamboo and wood are used in the manufacture of **rayon**. Wood and bamboo pulps are used for manufacturing paper. Every year, more than 10 million trees are cut to cater to our demand for virgin paper.

Many products such as tannins, gums, drugs, spices, insecticides, waxes, honey, ivory and hides are provided by flora and fauna of forests. Also they serve as main source of food products such as fruits, leaves, roots, and tubers of plants, and meats of forest animals.

DEFORESTATION OR OVER-EXPLOITATION OF FOREST

Man is cutting down trees in forests to get temporary benefits but this may lead to a tremendous loss in due course of time. The cutting down of forests is termed as **deforestation**. The tendency of deforestation is increasing day by day. The destruction of forests by natural activities and human activities has led to an overall deterioration of our environment and is posing a serious threat to the quality of life in future.

If a tree is cut down, energy stored in the wood and most of the nutrients of the system are lost. Such deforestation leaves a poor soil and cannot support agriculture for a long time, because the harvesting of the first few crops removes the remaining nutrients and renders it useless. Thus deforestation causes soil erosion.

The reduction of forests affects the rainfall and thereby restricts the availability of the most important natural resource—rainwater. In the natural forests, the roots bind the soil and about 90 per cent of the rainwater falling on the forests is retained either in humus or in that tissue. The forest thus acts as a soaking device and plays a vital role in the hydrological cycle. It has been estimated that in India, 60,000 million tons of topsoil is carried away annually by rainwater from a deforested area.

The current rate of deforestation is approximately one million hectares every year. International Union for the Conservation of Nature and Natural Resources says that India has lost nearly 2.5 million hectares of Mangrove forests over the last 80 years. The destruction of bamboo forests by the paper mills of Southern India has posed a considerable hardship for the local basket weavers; their earnings have fallen and now many are forced to steal their daily bamboo supply. Though the forest department controls 23% of India's total area, only about 10–12% has adequate tree cover. Cherrapunjee, the spot on earth once covered with lush subtropical forests, is today a barren area. The studies conducted at the International Institute for Sustainable Future, Mumbai, warn that the world's forest covers have been shrinking annually by almost the size of Nepal during the past 50 years.

Our former Prime Minister, Indira Gandhi, while addressing the Central Forestry Board in 1981, frankly admitted that only half of the official forestlands are under adequate tree cover. A report of the National Committee on Environmental Planning also clearly states that no more than about 12 per cent of the country's total land surface is under adequate tree cover. An aerial survey conducted by the Indian Space Research Organization in 1974 revealed that Anantapur district had no trees at all.

Case Study 1

Over-exploitation of Bamboo in Tamil Nadu

Thousands of poor villagers have suffered greatly because of the over-exploitation of bamboo livestock by the paper mills. They depend on bamboos for obtaining wood, for constructing houses, for manufacturing agricultural implements such as seed drills, etc., and for making mats and baskets (an important source of livelihood). It has been estimated that in Tamil Nadu alone, there are 23,000 people engaged in basket weaving, using bamboo as a means of survival. Probably a lakh of people would be dependent on these breadwinners.

Just about 10 years ago, the basket weavers of Kempanikanpalayam, a village near the town of Sathyamangalam in Tamil Nadu, used to go into their neighbouring forests, spend about half-a-day collecting bamboo, then come back and start weaving. The old people and others who could not go by themselves used to pay Rs. 50 per head load of bamboo. An average family would earn about Rs.100 per day. There was also an uninterrupted supply of the right quantity and quality of bamboo.

Now, some villagers specialized in collecting bamboo illegally are selling it to the weavers. These people have to travel nearly 25 miles to collect enough bamboo, and it takes about three days. Each head load now costs anything above Rs. 50; consequently, even with the prices of bamboo goods having risen, the weavers earn less than before. Moreover, the bamboo supply is no longer assured. Whenever forest officials capture the bamboo collectors, the supply may be stopped for weeks.

The destruction of forests across the Western and Eastern Ghats is a matter of major concern. Apart from illegal felling which remains a menace even in the southern states, several developmental projects like hydroelectric dams, the Kudremukh iron ore project, roads, extension of farm lands, etc., are also seriously eroding the forest base region.

Case Study 2

The Plywood Lobby

Upper Assam, especially the district of Dibrugarh, has native trees like the Hollong, which are the best in the world for plywood. The Hollong grows tall and straight for several feet. All its branches are concentrated at the top, and it is free of knots. Therefore, it is ideal for peeling and pressing into plywood sheets.

With the improvement in communications, roads and extension of the railway lines, the possibilities of a plywood industry in that region proved feasible. The Assam Government offered such attractive incentives that plywood millers operating in West Bengal closed their shops and moved to upper Assam. Today there are many plywood factories operating in the State, and they contribute a significant portion to the revenue of State. The plywood mills in Assam export decorative and commercial plywood and tea chest panels. In addition, they contribute to the Central and State exchequer, through central excise, sales tax, forest royalty, etc.

At present, the forests in Assam can meet only 22 per cent of the requirements of the State's plywood mills. The rest comes from Meghalaya, Arunachal Pradesh, and Nagaland and much through illegal felling in these States.

The plywood industries in Assam have been placed in difficulty because of the recent orders, which banned the movement of sawn timber outside their States, placed by the governments of Arunachal Pradesh and Nagaland. The pressure on Assam and Meghalaya forests is bound to increase. Illegal felling poses few hurdles for a timber contractor because the low-paid forest guards are easily bribed to turn the other way. The political influence of the plywood industry within the State ensures them of the Government's support in finding ways to meet their timber requirements—even at the cost of the forest.

Jhum cultivation in the two hill districts of Assam (North Cachar hills and Karbi Anglong), which have 40 per cent of the State's total forested area, is also contributing to deforestation. The firewood requirements of people, which are not met in any organized fashion, will leave the periphery of forests open to illegal felling.

Causes of Deforestation

The following are the major causes for the destruction and degradation of forests. The primary and secondary causes are clearly illustrated in Figure 2.2.

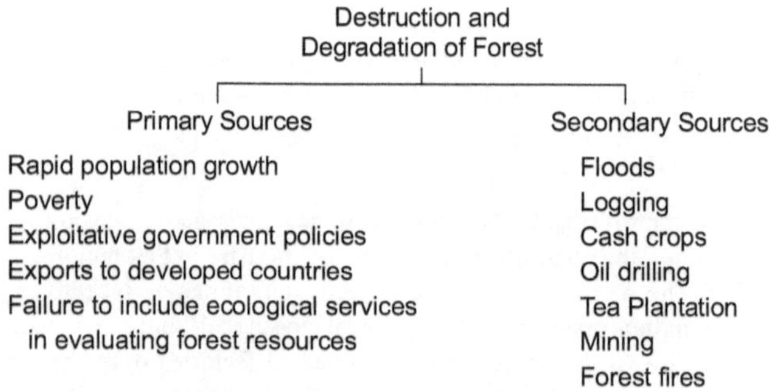

Figure 2.2   Causes of destruction and degradation of forests

Population explosion   Population explosion poses a grave threat to the environment. Vast areas of forest land are cleared of trees to reclaim land for human settlements like factories, agriculture, housing, roads, railway tracks, etc.

Forest fires   Forest fires may be due to natural calamities or human activities. They destroy the fully grown trees, which result in killing and scorching of the seeds, humus, ground flora and animal life. The forest fires are caused by any of the following reasons:

- ◇ Smouldering of the humus and organic matter forming a thick cover over the forest floor leading to ground fires.
- ◇ Dried twigs and leaves may catch fire, leading to surface fires.
- ◇ In densely populated forests, treetops may catch fire by the heat produced by constant rubbing against each other, leading to crown fires.

Grazing animals   Overgrazing by livestock has far reaching effects such as loss of porosity of soil, soil erosion and desertification of the previously fertile forest area.

Pest attack   The pests, e.g. insects, destroy the trees by eating up the leaves, by boring into the shoots and by spreading of the diseases.

Natural forces   Floods, storms, snow, lightening, etc. are the natural forces that damage the forests.

Mining and oil drilling   The finding of oil and coal within reserved forests means that further tracts of forests are cleared.

Effects of Deforestation

Forests are closely related with various factors such as climatic changes, biodiversity, wild animals, crops, land degradation, medicinal plants, etc. Figure 2.3 illustrates the effects of deforestation.

The deforestation of forests leads to

1. Increasing floods—flood damages alone now average Rs. 1000 crores every year.
2. Soil erosion.

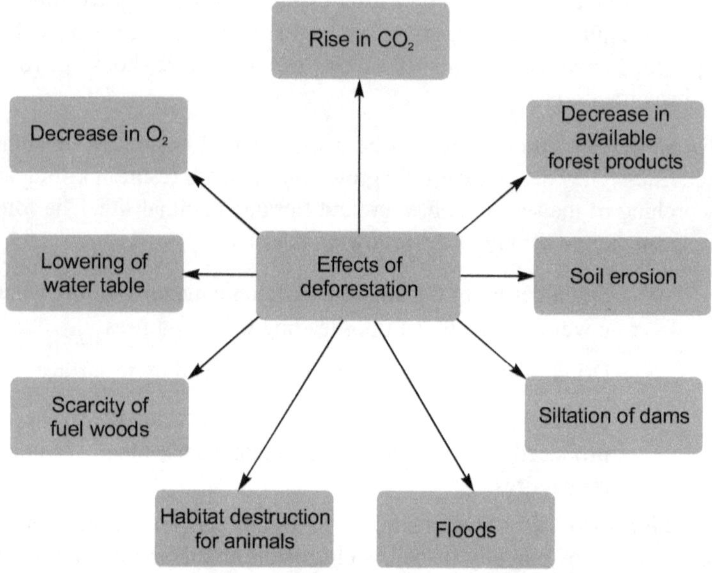

Figure 2.3   Effects of deforestation

3. Heavy siltation of dams built at an enormous expense and changes in microclimate.

4. Habitat destruction of wild animals.

5. Reduction in the oxygen liberated by plants through photosynthesis.

6. Increase in pollution due to burning of wood and due to reduction in carbon dioxide fixation by plants.

7. Decrease in the availability of forest products.

8. Scarcity of fuel wood, and deterioration in economy and quality of life of the people residing near forests.

9. Lowering of water table due to more run-off and increasing use of underground water increase the frequency of droughts.

10. Rise in carbon dioxide level has resulted in an increased thermal level of earth which in turn results in melting of ice caps and glaciers and in consequent flooding of coastal areas.

FOREST MANAGEMENT

The forest department is the custodian of 22,870 of forestlands, which constitute 17.584% of the geographical area in Tamil Nadu as against 33.33% required under National Forest Policy, 1988. Nearly half of the forest area is subjected to heavy degradation because of the biotic pressure. Various schemes and programmes of Government are aimed at restoring the degraded forest and expanding the forests outside the reserved forest area.

The FAO (Food and Agriculture Organization) has defined a forest as the land with a tree crown cover (or equivalent stocking level) of more than 10% and with an area of more than 0.5 hectare. The trees should be able to reach a minimum height of 5 m at maturity *in situ*. Forests are further subdivided into plantations and natural forests. Natural forests are the forests composed mainly of indigenous trees not deliberately planted. Plantations are forest stands established in the process of afforestation or reforestation by either planting, or seeding, or both.

Forests can develop wherever the average temperature is greater than 10°C in the warmest month and wherever the rainfall exceeds 200 mm annually. In any area having conditions above this range, there exists a variety of tree species grouped into a number of forest types that are determined by the specific conditions of the environment prevailing there, including the climate, soil, geology, and biotic activity. Forests can be broadly classified into types such as the taiga (consisting of pines, spruce, etc.), the mixed temperate forests (with both coniferous and deciduous trees), the temperate forests, the subtropical forests, the tropical forests, and the equatorial rainforests. The six major groups of forests in India are moist tropical, dry tropical, montane subtropical, montane temperate, subalpine, and alpine. These are again subdivided into 16 major types of forests. Consequently, forests were classified into four broad categories, namely forests for preserving the environmental stability, forests for providing timber supplies, forests for minor forest produces, and forests for pasturelands. While the first two categories were declared as reserve forests, the rest were designated as protected forests and were managed in the interests of the local communities.

India has a long history of traditional conservation and forest management practices. Under the British rule, the forest management systems were set in place mainly to exploit the forests. Nonetheless, there were some attempts to conserve forests and to meet the needs of local communities. The Indian National Forest Policy of 1894 provided the impetus to conserve India's wealth (forests) with the prime objectives of maintaining the environmental stability and of meeting the basic needs of the user-groups of fringe forests.

Soon after the independence, rapid development and progress were seen in large forest tracts fragmented by roads, canals, and townships. There was an increase in the exploitation of forest wealth. In 1950, the Government of India started the annual festival of tree planting called the **Vanamahotsav**. Gujarat was the first state to implement it. However, it was only in the 1970s that greater impetus was given to the conservation of India's forests and wildlife.

SOCIAL FORESTRY

The National Commission on Agriculture gave a serious thought to the problem of deforestation and recommended the introduction of "Social Forestry." India has been one of the first countries in the world to introduce the social forestry program to introduce trees in non-forested areas along roadsides, canals, and railway lines. Social forestry may be defined as an additional aid to wildlife conservation. According to K.M. Tewari, The President, Forest Research Institute, Dehradun, "Social forestry is a concept, a program and a mission that aims at ensuring ecological, economic and social security to the people, particularly to the rural masses especially by involving the beneficiaries right from the young stage to the harvesting stage".

Different Components of Social Forestry Programme

◇   Protection and afforestration of degraded forests.

◇   Creating of woodlots on community lands and government wastelands.

◇   Agro-forestry (trees along with agricultural crops) on marginal and submarginal farmlands.

◇   Control of erosion by planting trees or shrubs.

◇   Plantation of trees in and around habitation area and field boundaries.

◇   Plantation of trees in urban and industrial areas, for aesthetic purposes.

◇   Block and strip plantation along roadsides, canals, and rail lines.

CONSERVATION OF FORESTS

The following measures should be adopted to conserve the forests.

1.   The trees removed from the forest for any purpose must be replaced by new trees. Thus, tree felling should be matched by tree planting programme as early as possible.

2.   Afforestation should be done in the areas unfit for agriculture, such as along highways and riverbanks, and around playgrounds and parks. A special program of tree plantation called **Vanamahotsav** is held every year in our country. It should be made popular and effective.

3.   Maximum economy should be observed in the use of timber and fuel wood by minimizing the wastage.

4.   The use of firewood should be discouraged for cooking purposes and alternative sources of energy such as biomass, natural gas, etc., should be made available.

5.   Forests should be protected from fire. Modern equipments should be used to extinguish an accidental forest fire.

6.   Pests and diseases of forest trees should be controlled either by fumigation and aerial spray of fungicides or by biological control.

7.   Grazing of cattle in the forests should be discouraged.

8.   Modern methods of forest management should be adopted, e.g. use of irrigation, fertilizers, bacterial and mycorrhizal inoculation, disease and pest management, control of weeds, breeding of elite trees and tissue culture techniques.

9.   Improvement cutting and selective cutting should be done. Improvement cutting involves the removal of old dying trees, non-commercial trees, damaged trees, and diseased trees. Selective cutting involves cutting of mature timber trees and crowded trees.

Not all people own land, but every human being uses water. It is the lifeblood of our biosphere and is necessary for all forms of life. It connects us to one another, to other forms of life, and to the entire planet. Despite its importance to human survival and prosperity, water remains the most poorly managed resources on earth. It is squandered and polluted by industries, agricultural practices, sewage treatment plants and many other abuses. Water is treated as if it comes from an inexhaustible supply. This enhances the wastage and the pollution of this resource, for which we have no other substitute.

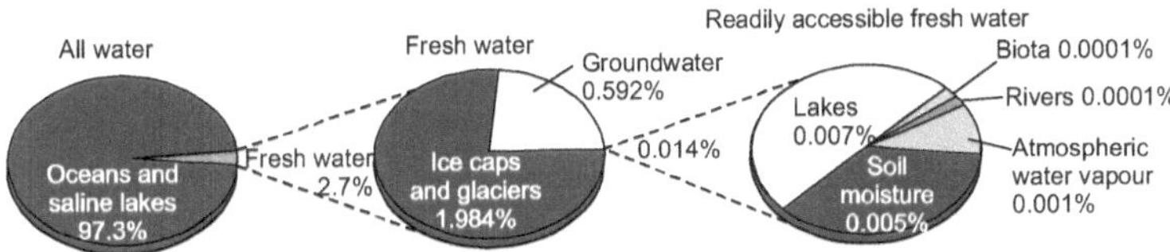

Figure 3.1 World's water supply for human use

About 2.7 per cent of the water on the earth is usable. The remaining 97.3 per cent is the saltwater that we cannot use unless it is desalinated. Of the 2.7 per cent that we can use, about three-fourths is frozen in the ice caps and glaciers, leaving just one quarter for our household uses (Figure 3.1). Thus, only about 0.014% of the earth's total volume of water is easily available to us as soil moisture, usable groundwater, water vapour, and lakes and streams. If the world's water supply is only 100 litres, our usable supply of fresh water would be only 0.014 litre, i.e., 2.5 teaspoons.

Canada, with only 0.5% of the world's population, has 20% of the world's usable freshwater, whereas China, with 21% of the world's people, has only 7% of the supply.

The available freshwater amounts to a generous supply. Moreover, this water is continuously collected, purified, recycled, and distributed in the solar-powered hydrological cycle. However, this works only as long as we either do not overload our water systems with slowly degradable and non-degradable wastes or do not withdraw water from underground supplies faster than it is replenished.

HYDROLOGICAL CYCLE OR WATER CYCLE

Natural ecosystems continually purify and recycle water as well as regulate its flow. The water cycle (Figure 3.2) consists of an alternation of evaporation and condensation/precipitation. Evaporation and transpiration are the purification stages of the water cycle. Precipitation is used to refer to any form of moisture such as rain, snow sleet and hail, which falls from the sky. If condensation occurs above the freezing point of water, drops of water form and fall as rain. If the condensation occurs at temperature below the freezing point, water molecules form crystals and fall as snowflakes.

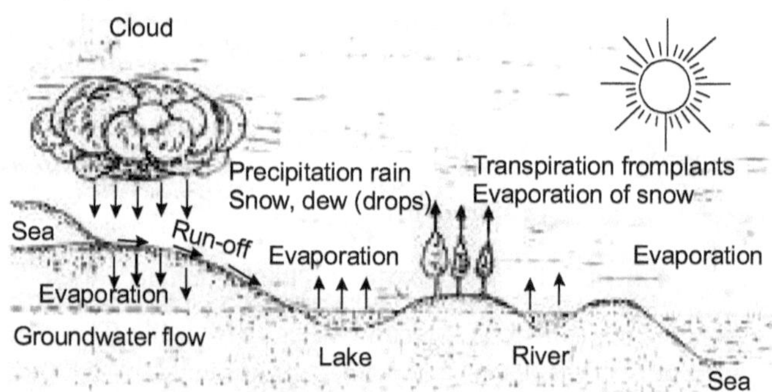

Figure 3.2   Water cycle

The amount of precipitation is the primary factor in determining the types of ecosystem that an area can support. The distribution of precipitation basically depends upon the pattern of heating and cooling of earth's atmosphere. The precipitate landing upon the ground becomes infiltration and run-off. The infiltration decreases with an increase in the run-off and increases with a decrease in the run-off. Run-off flows directly into streams. The water that is absorbed and held in the soil is called **capillary water**, which is available to plants. If it trickles down through longer cracks, pores, and spaces between particles and is being pulled by gravity, then it is called **gravitational water**. If it is accumulated above an impervious layer, it is called **groundwater**—the upper surface of which is the water table.

Thus, it connects all the ecosystems, i.e., from rainfall to moist soil, to springs, to streams and rivers and to the ocean. Humans, however, tend to carry on activities with little understanding of or regard for the natural water cycle. This threatens our livelihood and perhaps our existence. The following are the impacts of man-made activities on the soil and water cycle.

i.   Many streams that have a generally modest flow of clear water throughout the year are changed in character to become torrents of muddy water when it rains and to become dry channels when it does not.

ii.   Floods are frequent.

iii.   Rivers, reservoirs, and harbours are filling with sediments.

iv.   If the cost of environmental damage is not added to the cost of products, a market force that increases pollution will be generated.

USES OF WATER

Water is useful for domestic, industrial, agricultural and other purposes as described below.

◇   The high specific heat capacity of water helps to protect living organisms from temperature fluctuations. This property makes water an excellent coolant even for car engines, power plants, and heat-producing industrial processes.

◇   Water helps to distribute the heat throughout the planet and determines the climates of various areas.

◇   It carries dissolved nutrients into the tissues of living organisms, flushes waste products out of those tissues, serves as an all-purpose cleanser, and helps to remove and dilute the water-soluble wastes of civilization. (Water's superiority as a solvent also means that it can be easly polluted by water-soluble wastes.

◇   It helps to maintain a balance between acids and bases in cells, as measured by the pH of water solutions.

◇   It filters out the wavelengths of ultraviolet radiation that would harm aquatic organisms.

◇   The strong attractive forces between the molecules of water cause its surface to contract (high surface tension) and to adhere to and coat a solid.

◇   The combination of high surface tension, wetting ability and capillary force allow water to rise through a plant from the roots to the leaves.

◇   Because water expands upon freezing, it can break the pipes, crack a car's engine block (that is why we use antifreeze), break up the streets, and fracture the rocks (thus forming soil).

SOURCES OF WATER

There are two different sources of water.

1.   Surface water
2.   Groundwater

Surface Water

The precipitation that does not infiltrate the ground or return to the atmosphere by evaporation, including transpiration, is called as **surface run-off**. This run-off flows into streams, lakes, wetlands, estuaries, and reservoirs. About two-thirds of the world's annual run-off is lost by seasonal floods and is not available for human use. The remaining one-third is a reliable run-off, which we can generally account on as a stable source of water for years.

Groundwater

Some precipitation infiltrates the ground and percolates downward through voids like pores, fractures, crevices, and other spaces in soil and rocks. The water in these voids is called the groundwater. At or near the surface, the voids have little moisture in them. However, below some depth, i.e., in the zone of saturation, the voids are completely filled with water.

The water table is located at the top of the zone of saturation. It falls in dry weather and rises in wet weather.

Porous, water-saturated layers of sand, gravel or bedrock, one of through which the groundwater flows are called **aquifers**. They are the sources of drinking water. In the United States, water pumped from the aquifers supplies about 51% of the drinking water and 43% of the irrigation water. They are replenished naturally by natural recharge, where precipitation percolates downward through soil and rock, and sometimes (when water table is lowered) by lateral recharge where it is percolated laterally from nearby streams.

The groundwater normally moves from the points of high elevation and pressure to the points of lower elevation and pressure. This movement is quite slow and is thus typically released into streams, lakes, wetlands, or springs. The baseflow of streams and rivers, which is a sustained flow between storm events, is provided by groundwater. Thus if we disturb the hydrological cycle by removing groundwater faster than it is replenished, some nearby streams, lakes and wetlands can dry up.

GLOBAL WITHDRAWAL OF SURFACE WATER

According to a 2002 report, our global withdrawal of water had increased sevenfold (Figure 3.3) and per capita withdrawal had quadrupled. Uses of the withdrawn water vary from one region to another and from one country to another. Worldwide, about 69% of the water withdrawn each year from the rivers, lakes, and aquifers, is used to irrigate17% of the world's cropland arid to produce about 40% of the world's food. The industries use about 23% of the water withdrawn each year, and cities and residences use the remaining 8%. Agriculture and manufacturing processes use large amounts of water.

As a result, we are now withdrawing about 35% of the world's reliable run-off. We leave another 20% of this run-off into streams to transport goods by boats, to dilute pollution, and to sustain fisheries and wildlife. Thus we directly or indirectly are already using about 55% of the world's reliable run-off of surface water.

Because of increased population growth and economic development, global withdrawal rates of surface water could reach more than 70% of the reliable surface run-off by 2025. As countries press against the limits of available water, the possibility of armed conflicts over water supplies will increase.

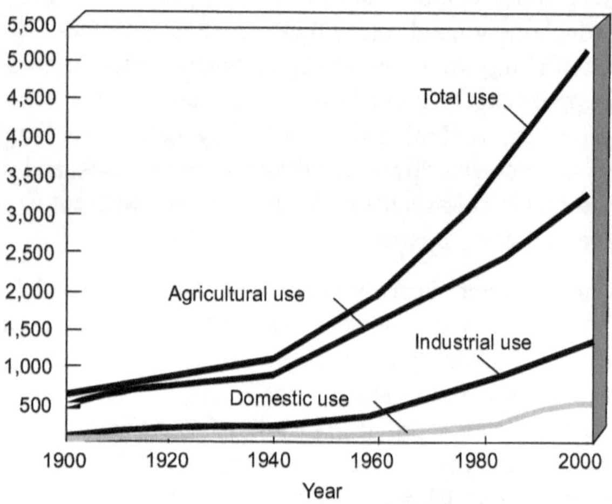

Figure 3.3　Global withdrawal of water

Some of the water withdrawn from a source may be returned to that source for reuse. Consumptive water use occurs when the water withdrawn becomes unavailable for reuse in the basin from which it was removed—mostly because of losses such as evaporation or contamination.

OVER-EXPLOITATION OF GROUNDWATER

The demands on groundwater are now increasing. The trend to irrigate from borewells is creating enormous new demands on groundwater. Groundwater is like a bank account. To maintain it for a long term, withdrawals should not exceed the deposits. According to World Resources Institute, water tables are falling in many areas of the world, as the rate of water pumping exceeds the rate of recharge from precipitation. If groundwater is withdrawn faster than it is recharged by infiltration of freshwater from surface, the water table will drop until dryness is reached. Not only the withdrawal rates of groundwater exceeding the recharge rates but also the recharge rates are reduced because of increased run-off and reduced infiltration. Figure 3.4 shows the exploitation of groundwater.

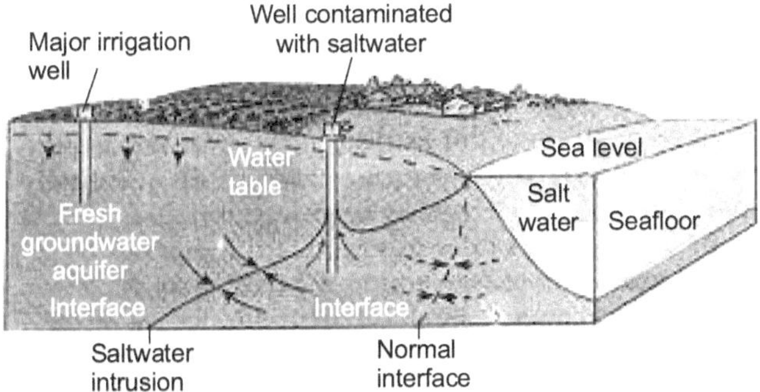

Figure 3.4   Exploitation of groundwater

Another problem resulting from groundwater removal is the saltwater intrusion. In the coastal regions, springs and seeps of outflowing groundwater may lie under ocean. As long as the water table is higher than the ocean level and the pressure in the aquifers is maintained, there is a net flow of freshwater into the ocean. Thus, wells near the ocean yield freshwater. The lower water table and/or rapid rates of groundwater removal may reduce the pressure in the aquifer, permitting the saltwater to push back into the aquifer. This is an especially serious problem in the coastal areas of Florida, California, South Carolina, and Texas. Among the major countries, India has over 50% of its area irrigated from groundwater, followed by the USA (43%), China (27%) and Pakistan (25%). That percentage can reach as much as 80% in developed countries with mild climate (Germany) and in arid countries (Saudi Arabia and Libya).

Exploitation of Water in India

In India, according to some estimates, 70–80 per cent of the production depends on groundwater irrigation. India greatly relies on extensive irrigation and it currently withdraws groundwater at twice the recharge rate, thus lowering the water table. Eventually, the shortfall could reduce the India's harvest by as much as 25 per cent. Since India well on the way to becoming the world's most populous nation already maintains an uneasy balance between production and consumption, the impacts could be disastrous. The limit of groundwater development for irrigation in the world's major **grain-producing (breadbasket)** countries may be reached in a few years before the end of the present decade in India.

Global Exploitation of Water

In the United States, groundwater is being withdrawn at four times its replacement rate. The most serious overdrafts are occurring in parts of the huge Ogallala Aquifer, extending from southern South Dakota to Central Texas and in parts of the arid south-western United States. Serious groundwater depletion is also taking place in California's water-short Central Valley, which supplies about half the country's vegetables and fruits.

Over-exploitation of groundwater resources in arid and semi-arid countries has been documented in recent studies. Aquifer depletion is a problem in Saudi Arabia, central and northern China, north-west and northern India, northern Africa (especially Libya and Tunisia), southern Europe, the Middle East, and parts of Mexico, Thailand, and Pakistan.

Table 3.1 Contribution of groundwater used for irrigation

Country	Per cent of groundwater as part of water resource used	Total irrigation water used (million m^3/year)
Saudi Arabia	96	15,310
Bangladesh	69	12,600
Tunisia	61	2,730
Jordan	55	740
India	53	460,000
Iran	50	64,160
Pakistan	34	150,600
Morocco	31	10,180
Mexico	27	61,200
China	18	407,800
South Africa	18	9,580
Nepal	12	28,700
Peru	11	16,300
Malaysia	8	9,700
Egypt	4	45,400
Mali	3	1,320
Indonesia	1	69,200

Groundwater development has been growing at an exceptional rate in recent decades. More reliable water delivery and declining extraction costs due to advances in technology and, in many instances, the government subsidies for power and pump installation have encouraged private investment in tubewells. For example, in India and Northern China, the area irrigated by groundwater rose from about 25% in the 1960s to well over 50% in the 1990s. Source of irrigation water varies widely between countries depending on the hydro-geological and climatic conditions, and the historical development of irrigation. Table 3.1 shows the contribution of groundwater used for irrigation for a set of countries that together represent about 150 million hectares or 56.6 per cent of the total irrigated area worldwide.

The recent development of groundwater has led to the over-exploitation of groundwater resources in some semi-arid regions where water tables have been falling at an alarming rate often one to three metres a year. These regions include some of the world's major grain-producing areas such as the Punjab and the North China plain.

Problems in Removal of Groundwater

i.   Overpumping of groundwater causes aquifer depletion. Sandra Postel, a water resource expert, says that about 480 million of the world's 6.3 billion people are being fed with grains produced with eventually unsustainable water mining from aquifers. This is expected to increase, as irrigated areas are expanded to help feed the 3.3 billion more people projected to join the ranks of humanity between 2000 and 2050. Thus overpumping the aquifers limit the future food production. In addition to this, it increases the gap between the rich and the poor in some areas.

ii. Groundwater removal causes the subsidence or sinking of land. The withdrawals have led to groundwater overdrafts and subsidence or sinking of land below aquifers in various parts of the United States.

iii. Groundwater removal causes salt extrusion of drinking water supplies. This is more common in coastal areas where the saltwater is pushed back into the aquifers. (See over-exploitation of groundwater.)

iv. Groundwater removal lowers the water table. As water tables drop, farmers must drill deeper wells, buy larger pumps to bring the water to the surface, and use more electricity.

v.   Groundwater removal increases the cost and energy consumption. Overdrafting groundwater also increases costs because more energy is needed to lift water (a heavy substance) from lower levels. Deeper wells can be drilled at great cost as water tables drop. But water pumped from such deep wells can be polluted because naturally occurring toxic elements such as arsenic, fluorine, and radioactive radon are more prevalent at deeper levels in the earth. Also, most of these substances dissolve in the hotter water found at these deeper levels.

Relying more on groundwater has both advantages and disadvantages (Table 3.2).

Table 3.2   Advantages and disadvantages of withdrawing groundwater

Advantages	Disadvantages
Good source of water for drinking and irrigation	Aquifer depletion from overpumping
Available year round	Sinking of land (subsidence) when water removed
Exists almost everywhere	Polluted aquifers unusable for decades of centuries
Renewable if not over pumped or contaminated	Saltwater intrusion into drinking water supplies near coastal areas
No evaporation losses	Reduced water flows into streams, lakes, estuaries and wetlands
Cheaper to extract than most surface waters	Increased cost, energy use, and contamination from deeper wells

Over-utilization and Pollution of Groundwater

Groundwater and surface-water pollution have increased. To raise the yield, farmers have increased inputs of herbicides, insecticides, fertilizers, and irrigation water on some crops. Many of these chemicals have percolated and accumulated to dangerous levels in the groundwater. The pollutants from upstream or nearby sewage treatment plants and industries can damage the productive zones which are fragile. Oil slicks can penetrate the zone with disastrous effects. Upstream dams that hold back the sediments can cut off nutrient supplies. Urban and industrial users may withdraw much of the freshwater upstream so that rivers run dry. Estuary zones (highly productive zone) are also subjected to dredging and filling that alter or destroy the habitats of plants and animals. Nowadays houses, factories, power plants, pipelines, oil refineries, hotels, warehouses, airports, and highways occupy millions acres of former estuarine zone.

WATER MANAGEMENT

Water management involves the protection of wetlands, estuaries, coastlines and rivers.

Wetlands

Wetlands are extremely valuable and productive habitats. They play an important role in regulating the stream flow. A study in Wisconsin indicated that in watersheds, where 40% of the land surface is a wetland, flood flows were 80% less than in areas without them, and the sediments in the streams were reduced by 90%.

Wetlands found along the freshwater stream, lakes, rivers and ponds are called **inland wetlands** whereas wetlands found along coastlines, including mangrove swamps, salt marches, and bays and lagoons are called **coastal wetlands**.

Wetlands hold water and allow an increased infiltration thus helping recharge the groundwater supplies. However, their usefulness is not always apparent and so they are often filled or dredged to make way for agriculture, housing, recreation, industry and garbage dumps.

Estuaries

Estuaries are critical for fish and shellfish reproduction. The estuaries zone has a high ecological productivity, because of the presence of nutrients that are washed away from the land. These nutrients support a number of aquatic organisms.

Coastal Areas

Beaches are also endangered by human activities. Though the sand is constantly replaced by the sediments carried by the rivers to the sea, dams that trap these sediments may halt the replenishment of coastal beaches.

Wild and Scenic Rivers

A river is a vital resource, but dammed and diverted to water-starved consumers it becomes a tragic symbol of poor planning and disregarded for the life forms that depend on its water. Perhaps the most serious deficiency in the regulation of water resources is the poor coordination between water management and economic development. The goal should be to use the river wisely and efficiently, minimizing the wastes and damages.

The use of groundwater for irrigation purposes is more recent in some countries. In Thailand, the excessive use of diesel pumps in the Chao Phraya and Mekong projects to extract groundwater for dry season cultivation has afftected the groundwater level.

CONSERVATION OF WATER RESOURCES

Air pollution, sedimentation, pollution of surface and groundwater and alteration of run-off can tragically reduce usable water supplies to intolerable levels. Much more thought, consideration and efforts need to be given to correct these problems at source. Preserving the integrity of water cycle is an important aspect of water conservation, more important than lowering the consumption of water. Unless we change the policies and priorities within 20 years, there won't be enough water for cities, households, the environment, or growing food. Water is not like oil. There is no substitute. If we continued to take it for granted, much of the earth is going to run short of water or food, or both.

Preventing and Reducing Flood Damage

In India, 90% of the annual precipitation falls between June and September. This downpour run-off so rapidly that most of it cannot be captured and used. The massive run-off also causes periodic flooding. There are number of effective ways to prevent or reduce flood damage. The vegetation can be replanted in disturbed areas to reduce this run-off. In urban areas, ponds can be built to retain the rainwater that will be released slowly to rivers.

Next to Bangladesh, India is the country verily affected by flood in the world and accounts for one-fifth of the global death count due to floods. About 40 million hectares, or nearly 1/8th of the India's geographical area, is flood-prone. The control measures consist mainly of construction of new embankments, drainage channels and afforestation. Flood forecasts are issued currently by Central Water Commission using conventional rainfall run-off models with an accuracy of around 65% to 70% having a warning time of six to twelve hours. An improved model for forecasting has been developed and was tested in lower Godavari basin in a pilot study titled "Spatial Flood Warning System." The deviation in the flood forecast from actual river flood has been within 15%.

Facing the Water Shortages or Droughts

Drought, overuse, wastage and overpopulation all contribute to the shortage of water. Water shortage is faced virtually by every nation in the world. In the developing countries, three out of every five people do not have access to clean and safe drinking water. Many countries are engaged in bitter rivalries over these precarious resources. The battle over the right for water is becoming more heated and entangled, as those who are living upstream and those who are living downstream both fight to protect their interests. Most of us will be using less water in the future whether we like it or not.

Drought is the most important weather-related natural disaster often aggravated by human action. Drought's beginning is subtle, its progress is insidious and its effects are devastating. Drought may start any time, last indefinitely and attain many degrees of severity. Since it affects very large areas for months and years, it has a serious impact on economy, leading to destruction of ecological resources, food shortages and starvation of millions of people.

As population, irrigation, and industrialization increase, water shortages in already water-short regions will intensify and heighten the regional tensions between and within countries. Global warming can increase the global rates of evaporation, shift precipitation patterns, and disrupt the water supplies and thus the food supplies. Some areas will get more precipitation and some less. Some river flows will change.

Drought prediction and management   Drought mitigation involves three phases, namely, preparedness phase, prevention phase and relief phase. In the case of drought preparedness, identification of drought-prone areas, and information on land use and land cover, wastelands, forest cover and soils are the pre-requisites. Space-borne multispectral measurements hold a great promise in providing such information.

Remote-sensing satellite data provide a major input for rainfall predictions, including long-term seasonal predictions and short-term predictions, which often help in the identification of developing drought conditions. Conventional methods of drought monitoring in the various States in India suffer from limitations with regard to timeliness, objectivity, reliability and adequacy. The focus has been on the assessment of agricultural drought conditions in terms of prevalence, relative severity level and persistence through the season. Such an exercise helps the decision-makers in initiating strategies for recovery by changing the cropping patterns and practices.

Several chronically drought-affected districts in India experience acute shortage of drinking and irrigation water. To address this issue, a nationwide project titled Integrated Mission for Sustainable Development (IMSD) was taken up in collaboration with other Department of Space Centres and State Remote Sensing Applications Centres. The project essentially aims at generating locale-specific action plan for development of land and water resources on a micro-watershed basis in drought-prone areas of the country using data from Indian remote-sensing satellites. In the first phase, 175 districts covering 84 million hectares have been covered.

Causes of water scarcity   The following are the main causes of water scarcity.

1. The water sources are used to the maximum level in order to meet the needs of the growing population.

2. The increasing buildings, tar and cement roads occupy the open space and prevent the percolation of natural rainwater into the earth.

3. Rainfall is the only source for all the available water. The failure of monsoon rains brings about increased water scarcity.

4. Usage of underground water becomes more during poor rainfall.

5. Excessive pumping of groundwater reduces the water level and the wells become dry.

The three factors that cause water scarcity are 1) a dry climate, 2) desiccation, and 3) water stress. During dry climate the precipitation is at least 70% lower and evaporation is higher than the normal. The term desiccation refers to the drying of exposed soil because of the activities such as deforestation and overgrazing by livestock. The low per capita availability of water caused by increasing numbers of people relying on limited run-off levels is known as **water stress**.

A country is said to be water stressed when the volume of its reliable run-off per person drops below about 1,700 cubic metres per year. A country suffers from water scarcity when per capita water availability falls below 1,000 cubic metres per year. Water stress and water scarcity if prolonged will lead to drought.

According to a 2002 report by the United Nations, about 500 million people live in countries that are water-scarce or water-stressed. By 2025, it may increase up to 2.4–3.4 billion people in water-scarce or water-stressed river basins, particularly in North Africa and west Asia. According to a former Environmental Minister of China, the freshwater supplies are capable of supporting only about half of its current population. In 2002, World Bank officials said that dwindling water supplies in various parts of the world would be a major factor that inhibits the economic growth and development. Also in the same year, the International Food Policy Research Institute and the International Water Management Institute used computer modelling to project that, by 2025, water scarcity will cause annual global losses of 350 million metric tons (385 million tons) of food production—slightly more than the entire current US grain crop.

In water-short rural areas in developing countries, each day many women and children must walk a long distance, carrying heavy jars or cans to get a meagre, and sometimes contaminated, supply of water. Since the 1970s, water stress and scarcity intensified by prolonged drought have killed more than 24,000 people per year and created millions of environmental refugees.

Industrial countries of the world with only one-third of the population have most of the available fresh water. In the high-consumption countries with rich resources and with a highly developed technical infrastructure, the many ways of water use and reuse may more or less be sufficient to curb further growth in supply. However, in other regions, water availability is critical to any further development above the present unsatisfactory low level and even to the mere survival of communities or to meet their growing demand due to increase in population. More than 1 billion people including one-third of the population of China and India live in arid regions facing water stress. In these regions, man cannot forgo the contribution to be made by dams and reservoirs for the harnessing of water resources in addition to all other solutions.

Constructing Dams and Reservoirs

Dams and reservoirs increase the supply of water, control floods, generate electricity and increase certain forms of recreation. Because the economic and environmental costs of dams and reservoirs are so high, non-destructive and less costly measures should be used wherever possible.

Dams—benefits and problems   Large dams and reservoirs have both benefits and drawbacks. Their main purpose is to capture and store the run-off and to release the same when required. It is needed for controlling floods, producing hydroelectric power, and supplying water for irrigation and for towns and cities. Reservoirs also provide recreational activities such as swimming, fishing, and boating. The dams have increased the annual reliable run-off available for human use by nearly one-third.

The fact is that more than 40,000 large dams that measure more than 15 metres in height are currently obstructing the world's rivers, whose reservoirs cover more than 400,000 square kilometres of land—an area larger than the combined surfaces of the United Kingdom, Belgium, The Netherlands and Austria. These reservoirs have inundated millions of hectares of forests, particularly in the tropics, of which many were not even logged and trees were left to slowly rot. Being displaced by the dams, the farmers have had to clear forests in other areas in order to grow their crops and build their homes. Therefore, dams have resulted in the deforestation. Additionally, dams imply road building, thus allowing access to previously remote areas by loggers and "developers", further enhancing the deforestation processes.

Case Study 3

Dams and their Effects on Forest

Large dams are generally constructed in areas where no modern developments have been initiated. More and more dams are constructed in environmentally sensitive sites such as in the Himalayas or in the heavily forested areas.

The forests are also being cleared for improving the approach roads, offices, and residential quarters and for storing the construction materials of a dam. With a reduction in forest cover along with the increasing population, the pressure on the remaining forests increases. Their need for firewood leads to further denudation. Thus, the construction of dam has several effects. At the time of planning of dam construction, the cost benefit calculations are not made genuinely; the forest loss and the recurring loss in terms of timber and firewood yield per year are grossly undervalued; and the social costs of degradation of environment are never taken into the account.

Dams and their Effects on Tribal People

The construction of several large dams has evicted thousands of families and causes their resettlement at some other site. One of the first major eviction operations is the evacuation of the town of Bilaspur in Himachal Pradesh, which was submerged by the Bhakra dam reservoir. Seven thousand cities of this town having modern amenities like colleges, hospitals, markets, and playgrounds were displaced. A new township was created on the nearby high ground and the displaced persons were given cash to build their new houses. Today, the new Bilaspur is a flourishing town. The process of relocation will mean further deforestration.

Tribal population continues to suffer at the hands of non-tribal populations. In the hilly district of Wynad in north Kerala, forest encroachers in association with forest officials have acquired nearly 80 per cent of the land belonging to the local tribals. While only 20 per cent of the tribal in the district were classified as landless agricultural labourers 30 years ago, today they constitute about 80 per cent. Legal enactments to remedy this land alienation of tribals have met with powerful opposition from vested interest.

In addition to deforestration, dams also influenced major environmental changes at both the dam site and in the entire river basin. Not only are the best agricultural soils flooded by the reservoir, but also the river's flora and fauna begins to disappear, with strong impacts on people who are dependent on those resources. At the same time, dams imply a number of health hazards, starting with diseases introduced by the thousands of workers who were brought to build the dam (including AIDS, syphilis, tuberculosis, measles and others) and ending with diseases related to the reservoir itself (malaria, schistosomiasis, river blindness, etc.).

A series of dams on a river, especially in arid areas, can reduce downstream flow to a trickle and prevent it from reaching the sea as a part of the hydrologic cycle. According to the World Commission on Water in the 21st century, half of the world's major rivers are going dry part of the year or are seriously polluted. In addition to threatening the water supplies for 500 million people, this engineering approach to river management often impairs some of the important ecological and economic services that the rivers provide.

So far, in too many cases, dam construction has resulted in widespread human rights violations. The people belonging to that locality have persistently resisted the destruction of their homelands and their forced "resettlement". As a result, they have faced different types of repression, ranging from physical and legal threats to mass murders, such as in the case of the Chixoy dam in Guatemala. But resistance, consciousness and solidarity have grown. These people are now able to organize themselves and to establish local, national and international alliances with other concerned organizations, e.g. the Narmada Bachao Andolan Movement in India.

Case Study 4

China's Three Gorges Dam

China's Three Gorges Project on the mountainous upper reaches of the Yangtze River will be the world's largest hydroelectric dam and reservoir. The 2-km-long dam is supposed to be completed by 2013 at a cost of at least $25 billion.

About 1.9 million people will be relocated behind the dam. When completed, the dam will have the electrical output of 20 large coal-burning or nuclear power plants. It will also help hold back the Yangtze River's floodwaters, which have killed more than 500,000 people during the past. Table 3.2 lists the major advantages and disadvantages of this huge dam and reservoir.

Advantages	Disadvantages
Generates 10% of China's electricity	Floods large areas of cropland and forests
Reduces dependence on coal	Displaces 1.9 million people
Reduces air pollution	Increases water pollution because of reduced water flow
Reduces CO_2 emissions	Reduces deposits of nutrient-rich sediments below dam
Reduces chances of downstream flooding for 15 million people	Increases saltwater intrusion into drinking water near mouth of river because of decreased water flow
Reduces river silting below dam by eroded soil	Disrupts spawning and migration of some fishes below dam
Increases irrigation water for cropland below dam	High cost

Case Study 5

Tehri Dam and its Impacts

Tehri Dam is 1.5 km downstream from the point of confluence of the Bhagirathi and Bhilangana rivers, and is about 80 km north of Rishikesh in Uttar Pradesh. It is formed with rocks and rubble and is technically termed as a rock-fill dam. It is 260.5 metres high, making it the fifth highest of its kind in the world. It is constructed to serve for dual purposes. First, to generate hydroelectric power with an installed capacity of 2000 MW, and the second is to provide water for irrigating about 668,000 hectares of land.

But geo-scientists and environmentalists suggest that the size of the dam and its location could trigger an earthquake in a seismically active area. Even most of the solid rocks found in the river gorge were found to contain open joints, weak zones, fractures, fissures, cleavages and other tectonic features. When the dam overflows, there will be excessive water seepage that will weaken the structure. As a result, the dam could collapse just as the rock-filled Teton dam broke in Idaho, USA, in June 1975. The location of this dam is across the narrow gorge. The pressure of 3.2 billion tons of water impounded in the reservoir could trigger off an earthquake bigger than that of in Koyna, Maharashtra, which led to a great loss of lives and properties.

Another important hazard is the heavy rate of siltation of the snow-fed rivers such as Bhagirathi and the Bhilangana, which carry large quantities of fine silt and boulders and create stresses such that the dam could collapse. The siltation rate of these rivers has been assumed to be 110-acre-feet in 100 square miles per year, giving a life span of about a hundred years for the Tehri dam. However, similar assumptions about other large dams have proved to be vastly conservative making their actual life spans much less.

The building materials needed for its construction, namely, granite, quartzite, sandstone, etc., have created additional problems. The materials are not locally available. Construction of roads and building activity have loosened the rocks, caused landslides and worsened the soil erosion and siltation problems.

Over 32,000 people were affected in Tehri town and in 92 villages. The process of relocation will mean further deforestation. The controversies over the benefits and hazards of the Tehri dam reveal that its environmental impact has simply not been studied thoroughly.

Water Diversion Projects

One way to increase the natural supply of water to a particular area is to divert water from areas where it is in plentiful supply. Large-scale water diversion projects are controversial and expensive. A river or other body of water is damaged when major portion of its water is diverted. Pollutants, which would have been diluted in the normal river flow, reach higher concentrations when much of the flow has been removed.

Dams alter the river ecosystems downstream. For example, the Glen Canyon dam (1963) affected the Colorado River. The level of Mono Lake in eastern California was reduced by 14 metres, and salinity increased dramatically since the surface water was diverted to Los Angles 442 km away. Aral Sea in Kazkston is suffering from the same problem as Mono Lake. Its storage area is declined by 50% and total volume by 75%. Opponents of water diversion projects contend that serious water conservation efforts would eliminate the need for additional large-scale diversion.

Case Study 6

Advantages and Disadvantages of Large-scale Water Transfer

The Aral Sea Disaster Tunnels, aqueducts, and underground pipes can transfer the run-off collected by dams and reservoirs from water-rich areas to water-poor areas. Although such transfers can bring water to water-short areas, they also create environmental problems. Indeed, most of the world's dam projects and large-scale water transfers illustrate the important ecological principle that you cannot do lust one thing.

An example of this principle in action is the shrinking of the Aral Sea. It is a result of a large-scale water transfer project in an area of the former Soviet Union with the driest climate in central Asia. Since 1960, enormous amounts of irrigation water have been diverted from the inland Aral Sea and its two feeder rivers to create one of the world's largest irrigated areas. The world's longest irrigation canal stretches over 1,300 kilometers.

This large-scale water-diversion project (coupled with droughts) and high evaporation rates in this area's hot dry climate have caused a regional, ecological, economic, and health disaster. Here are some of the problems caused by this massive water-diversion project.

1. Salinity of the sea has tripled.

2. Surface area of the sea has decreased by 54%

3. Volume of the sea has decreased by 75%

The shrinkage of the Aral lake has caused one of the world's worst salinization problems. The wind picks up the salty dust that encrusts the lake's now-exposed bed and blows it onto fields as far as 300 kilometres (190 miles) away. As the salt spreads, it pollutes water and kills wildlife, crops, and other vegetation. Aral Sea dust settling on glaciers in the Himalayas is causing them to melt at a faster-than-normal rate.

This process has converted about 36,000 square kilometres (14,000 square miles) of former lake bottom to a human-made desert covered with glistening white salt. It has also caused the presumed extinction of 20 of the area's 24 native fish species from the increased salt concentration. This has devastated the area's fishing industry, which once provided work for more than 60,000 people. Fishing villages and boats once on the sea's coastline now are in the middle of a salt desert and have been abandoned.

About 85% of the wetlands in the area have been eliminated, which along with increased pollution has greatly reduced waterfowl populations. Roughly half the area's bird and mammal species have disappeared.

CONFLICTS OVER WATER

Water plays a central role in assuring an adequate food supply. Agriculture is the largest consumer of water, using an average of 80 per cent of total water consumption in the developing countries. As population swells and incomes grow, the demand for water for residential and industrial purposes increases rapidly. With water becoming increasingly scarce and development of new sources of water becoming very costly for many already capital-short countries, more national conflicts are expected to arise—industry, urban centres, and agriculture set against each other. To solve these national water tensions, countries tend to look beyond their borders for wider reign over water basins that they share with other countries. Hence, national water scarcity could escalate the existing tensions between nations and lead to flare-ups of long-standing international water conflicts. Unless this changes, the world may be faced with another polarizing force to replace the cold war.

But water scarcity and water conflicts can be prevented if water conservation or **water use efficiency** is increased in agriculture. If policies like subsidies for water are changed, there may be enough water for agriculture and to the fuel industry and to the urban centres of the future. Without such changes, however, water wars could erupt.

Conflicts are brewing now over rivers and river basins shared by many countries around the world. It is expected that more than 200 bodies of water are shared by two or more countries.

In Middle East

The conflicts over water are erupting throughout the Middle East, from the watersheds of the Nile to the Tigris and Euphrates Rivers. Middle East region is running out of water. And the people, who have built their lives and livelihoods on a reliable source of freshwater are facing the shortage of this vital resource, impinge on all aspects of the tenuous relationships which have developed over the years between nations, between economic sectors, and between individuals and their environment.

In South Asia

Likewise, a tussle is simmering in South Asia's Ganges—Brahmaputra Basin—where Bangladesh, India, and Nepal dispute for the best uses of water. India and Nepal want to exploit the basin's huge potential of generating hydroelectric power, whereas Bangladesh wants the water to be managed in such a way as to minimize flooding during monsoon months and to reduce water shortages during dry months. Of equal concern are the water conflicts between states in India that share river basins, such as Karnataka and Tamil Nadu, which border the Cauvery River. In the other parts of India, people have been killed in these conflicts over water. There are water shortages during the dry season in every major city in South Asia. During the dry season, city pipes are often empty, creating situations where water must be delivered by trucks to mobs of desperate people.

Water and International Conflicts

River basins and groundwater aquifers that cross international boundaries present increased challenges to effective water management because hydrologic needs are often overwhelmed by political considerations. While the potential for paralysing disputes is especially high in these transnational basins, the record of violence is actually greater within a nation's boundaries.

The scarcity of water, required by humans and other ecosystems, leads to intense political pressures, often referred to as "water stress." As a consequence, competition for water resources has contributed to tensions around the globe. Water demands are increasing, groundwater levels are dropping, surface-water supplies are increasingly contaminated, and delivery and treatment infrastructure is ageing.

These tensions have spilled into violence on occasion, mostly at the international level, and generally among ethnic, religious, or tribal groups, water-use sectors, or states/provinces. Examples of internal water conflicts range from interstate violence and death along the Cauvery River in India.

While these disputes can and do occur at the sub-national level, the human security issue is subtler and more pervasive than violent conflict. As the quality of water degrades and the quantity diminishes over time, tensions can spill across boundaries. The overall effect on the stability of a region will be unsettling.

Background on International Waters

There are 261 watersheds and countless aquifers that cross the political boundaries of two or more countries. International basins cover 45.3 per cent of the land surface of the earth, affect about 40 per cent of the world's population, and account for approximately 60 per cent of global river flow. These basins have certain characteristics that make their management especially difficult, most notably the tendency for regional politics to exacerbate the already difficult task of understanding and managing complex natural systems. The disparities between riparian nations—whether in economic development, or in infrastructural capacity, or in political orientation—further complicate the international water resources management. As a consequence, development projects, treaties, and institutions are regularly observed as inefficient at best and, occasionally, as a source of new tension themselves.

Development, Crisis, and Conflict Resolution

A general pattern has emerged for international basins over time. Riparians of an international basin first unilaterally implemented water development projects on water within their territory in an attempt to avoid the political intricacies of the shared resource. At some point, one of the riparians, generally the regional power (either upstream riparian or country with the most military, political, or economic strength), will implement a project that impacts at least one of its neighbours. The project might aim either to continuously meet the existing demand in the face of decreasing relative water availability, such as Egypt's plans for a high dam on the Nile, Indian diversions of the Ganges to protect the port of Calcutta, etc., or to meet new needs reflecting new agricultural policy, such as Turkey's GAP Project on the Euphrates. A project that impacts one's neighbours can become a flashpoint in the absence of relations or institutions conducive to conflict resolution, heightening tensions and regional instability, and requiring years or more commonly decades, to resolve.

The problem gets worse as the dispute gains in intensity. Some ecosystems, such as the lower Nile, the lower Jordan, and the tributaries of the Aral Sea, have been allowed to deteriorate because authoritative states did not cooperate to protect them. They have effectively been written off to the vagaries of human intractability.

MINERALS

Mineral resources are the naturally occurring materials in or on the earth's crust that can be extracted and processed into useful materials at an affordable cost. The main source of minerals is the earth's crust, as it is made up of minerals and rocks. A rock may contain more than one mineral. The minerals may consist of a single element, e.g. gold, silver, diamond (carbon), and sulphur, or consist of various combinations of elements, e.g. salt, mica and quartz. More than 2000 minerals occur as inorganic compounds. Over millions to billions of years, the earth's internal and external geologic processes have produced numerous non-fuel mineral resources and energy resources. Because they take so long to produce, they are classified as non-renewable resources.

USES OF MINERALS

Metals have three properties that make them valuable.

1. They are strong.
2. They can be melted and cast into any shape that a mould can provide.
3. They conduct electricity very well.

In recent years, scientists have developed a variety of new materials that can replace metals for variety of uses.

LIFE CYCLE OF METAL RESOURCE

When metals are to be first exploited, ores have to be dug out of the ground with primitive tools. It is possible to do this because rich, concentrated and surface deposits are available. Today most of these readily accessible deposits have been exhausted, but machinery is available to extract ores from deep underground and from harsh environments. The typical life cycle of a metal resource is depicted in Figure 4.1. It begins with the extracting of ore from the earth's crust. The processing of ore to extract the metals like copper and gold is called **mining**. The ore typically has two components. One is the mineral containing the desired metal and the other the waste material called **gangue** (pronounced "gang"). The gangue is removed from the ore. Most of the ores consist of one or more compounds of the desired metal. After gangue has been removed, smelting is used to separate the metal from the other elements in the ore mineral.

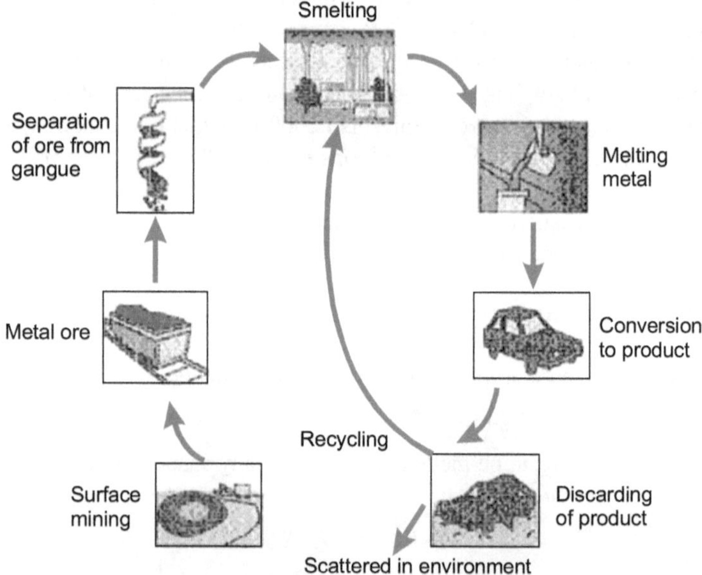

Figure 4.1   Life cycle of a metal resource

The metal thus obtained is melted and converted to a desired product. The cycle ends in the discarding of the product. Thus the metal is recycled for later use.

In modern times, mining operations confirm to a great variety of environmental regulations. The objective of these laws is to restore the land to its original condition where the mining operations are complete. In order to achieve this goal, biologists, ecologists, and soil scientists must study the ecosystems in their undisturbed states, and then must work together with engineers to direct the reclamation which is perforated by heavy equipment operators and their helpers.

ENVIRONMENTAL EFFECTS OF EXTRACTING AND USING MINERAL

Mining implies selection, selection in turn implies rejection of waste and the very process of selection may produce smoke, dust, noises and other undesirable effects. Mining, in short, contains inevitably the seeds of pollution; unchecked, they will grow and spread widely or sparsely as the case may be. The mining, processing, and use of mineral resources make enormous amounts of energy and often cause land disturbance, soil erosion, and air and water pollution. The environmental impacts of mining could be broadly classified into five categories, namely water pollution, air pollution, despoliation of land, deforestation, and noise and ground vibration problems.

Air Pollution

The addition of gaseous pollutants to the atmosphere and the emission of dust particulates during mining are the two factors involved in air pollution. The sources of gaseous pollutants, which include sulphur dioxide, oxides of nitrogen and carbon monoxide, are mainly from the waste dumps. The dust particulates on the other hand are produced during ore handling, blasting and transportation operations. Mine fires, a serious problem associated with the underground mining and solid waste disposal, are also, to a great extent, responsible for polluting the atmosphere. Without effective pollution control equipment, smelters emit enormous quantities of air pollutants, which damage vegetation and soils in the surrounding area. The machinery, trucks and blasting operations in a surface mine pollute the environment with air-pollutants, dust and noise.

Water Pollution

The mining activities affect the hydrology of the area in many ways. Sometimes, the course and the rate of flow of stream changes affect the agriculture and flora and fauna of the area. Lowering of water table is another phenomenon resulting from mining activities. The process of removing the gangue from the ores produces piles of waste called **tailings.** Particles of toxic metals blown by the wind or leached from tailings by rainfall can contaminate the surface water and the groundwater. Smelters also cause water pollution and produce liquid and solid hazardous wastes that must be disposed off safely.

Surface mining often leads to acid mine drainage and silt run-off which degrades quality of water. Acid mine drainage is an important source of increased acidity in natural waters. Its main constituents are sulphuric acid and iron compounds formed as a result of water, air and pyrite and are drained into the natural water sources (Figure 4.2). This contaminates the water supplies, and can destroy some forms of aquatic life. The discharge of acid mine water into streams/water pollution and the addition of toxic substances like nickel, fluorine, and other elements in the mining operations are the causes of water pollution.

Figure 4.2   Iron hydroxide precipitate stains a stream receiving acid drainage from surface coal mining

Despoliation of Land

Subsidence of land, due to underground mining and large-scale excavations, in surface mining, are the basic causes of land despoliation in course of the mining activities. Coal is mined either by surface or underground methods. Irrespective of the process, coal mining requires a lot of land area especially the open cast mining. For surface or open-cast mining, the ground is cleared of vegetation and the soil is removed resulting in the soil erosion and disfiguring of the land considerably. The void formed by such a removal of top mineral-rich soil is filled with broken overburdened rock. A part of the void is left as such. Greater the depth of the open pit the greater is the disfigurement of land. This results in scarring and disruption of the land surface. Landslides are another result of surface mining.

Land degradation due to open-cast mining   Usually surface mining methods produce large excavations, and the resulting overburden and mineral wastes are spread over large areas, making them unusable (Figure 4.3). Open-cast mining also leads to rapid erosion of land because of slope stability problems and denudation of vegetation. Detrimental effects of soil erosion are ultimately felt in silting and degradation of streams and water bodies.

Figure 4.3   Chuquicamata, the largest and second deepest open pit copper mine in the world, Chile

Another problem associated with mining, which results in despoliation of land, is the problem of solid waste disposal. The waste dumps, which are created as a result of open-cast and underground mining could lead to slips or erosion if not constructed and treated properly.

Subsidence of land The phenomenon of surface subsidence due to underground mining is well known. This subsidence can cause the houses to tilt, sewer lines to crack, gas mains to break, and groundwater systems to be disrupted (Figure 4.4). This not only affects the buildings and the surface drainage pattern but sometimes even highways, bridges, water and gas lines may be sheared, twisted and broken by the strains and slope changes resulting due to subsidence.

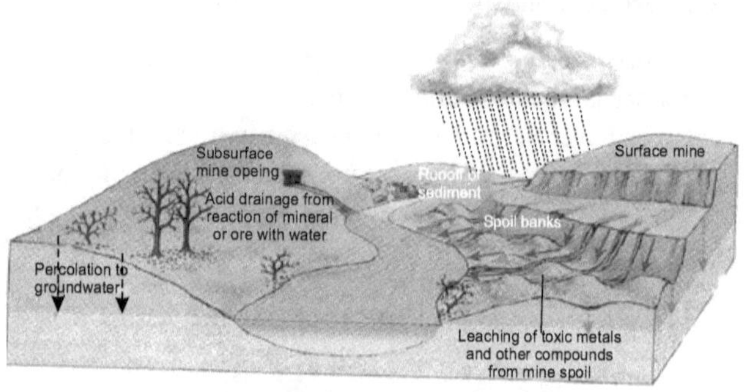

Figure 4.4 Mining

Deforestation

In many cases, the valuable ore bodies are located in either thick forests or adjacent to it. During the construction of infrastructure and pit site, the trees and vegetation are destroyed. In addition, the spreading of toxic overburden and dirt destroys the vegetation. Massive deforestation drastically changes the climatic conditions like rainfall, temperature and humidity. This may also result in the death of wild animals and birds living in that forest.

Noise and Ground Vibration Problems

With the introduction of sophisticated machinery into the mining operations, the noise problem, which was never considered to be serious in the past, also needs immediate attention. It has been established that noise interferes with speech communication, and causes annoyance and distraction. With the increase in the scale of mining operations, heavy planting has to be resorted to in the large open-cast mines. This results in ground vibrations, which may cause damage to the structures in the vicinity besides causing irritation to the people living nearby. Besides the above direct environmental impacts, mining operations may also cause adverse social and visual effects.

Effects of Mining on Human Health

Health and safety of mining is greatly at risk especially in underground mining. Explosions, fires, and cave-ins are common coalmine accidents resulting in innumerable deaths and casualities. Pneumoconiosis commonly known as the black lung disease is a major hazard to underground mining. In this disease, inhalation of dust by the miners impairs respiration by damaging the lung tissue leading to disability and often death, as there is no cure known for this disease. This disease can be prevented by adopting better techniques of mining like water spraying and better ventilation that considerably lowers dust pollution .

Mining can also result in the emission of toxic chemicals into the atmosphere. In the United States, the mining industry produces more toxic emissions than any other industry (typically accounting for almost half such emissions). Toxin-laced mining wastes can be blown or deposited elsewhere by wind- or water-caused erosion. Finally, some forms of wildlife can be exposed to the toxic mining wastes stored in holding ponds and to the leakage of toxic wastes from such ponds.

The economic activities of men result in ecological deterioration. The environmental damages should be arrested immediately to protect and improve the environmental quality. If proper technologies are adopted for carrying out economic activities such as mining and blasting, etc., it will definitely increase the output without increasing the environmental problems. For example, some companies are using improved technology to reduce pollution from smelting, to lower production costs, to save costly clean up bills, and to decrease liability for damages.

Case Study 7

Health Impacts of Domestic Coal Combustion in China

Domestic coal combustion is the primary source of energy for hundreds of millions of people worldwide. In many developing countries such as China, unventilated stoves are commonly used. Emissions from the unvented coal fires are known to cause a variety of health problems that affect tens of millions of people in China alone. There are serious health issues in China. The Institute of Geochemistry (IGC) and Chinese Academy of Sciences work together on the characterization of mineralized coals in Guizhou Province in southern China. The use of these coals in unvented stoves is believed to have caused several thousand cases of severe arsenic poisoning and as many as 10 million cases of fluorine poisoning.

In addition, it is possible that mobilization of mercury, thallium, antimony, and other toxic elements from these coals may have affected the health of people in this region. Moreover, several cases of selenium poisoning in this area have been attributed to the use of coaly shale.

Some of the coal samples from the Guizhou Province have resulted in as much as 35,000 parts per million (ppm) arsenic. Several samples have more than 50 ppm mercury and more than 200 ppm antimony. These values are about 1,000 times the concentration found in most other coals.

BIOMINING

It is the process of microbiological ore processing to extract the metals like copper and gold. Can we use microbes to mine metals? Since 1958, the copper industry has been using natural strains of the bacterium *Thiobacillus ferroxidans* to remove copper from low-grade copper ore. This technique is increasingly applied for low-grade ores, as high-grade ores are depleted. Currently, more than 30% of all copper produced worldwide comes from such biomining. If at times the naturally occurring bacteria cannot be found to extract a particular metal, genetic engineering techniques could be used to produce such bacteria.

However, microbiological ore processing is slow. It can take decades to remove the same amount of material that conventional methods can remove within months or years. To date, biological mining methods are economically feasible only with low-grade ores such as gold and copper for which conventional techniques are too expensive.

EXPLOITATION OF MINERALS

As long as we continue to live in a civilized society, metals will continue to be valuable. Moreover, there are various trends that will lead to an ever-increasing demand. How long will our mineral reserves last? Of course, some minerals are more abundant than others. However, future availability depends, in part, on the geology of the mineral in question as well as on its abundance. Once the richest deposits of some ores are exploited, there is a sharp decline in both the concentration of ore bodies and the total quantity of ore available.

Case Study 8

ILLEGAL MINING IN CHINA

About 10,000 gold miners are being given their marching orders in north China's Shanxi province in a move to shut down illegal mines. Nearly 40 people died in an explosion at a privately run gold mine in Yixingzhai, Fanzhi County, in June 2004. Thousands of miners' wives have lost their husbands.

China's huge mining industry has a terrible safety record with more than 3,000 deaths reported in the past six months alone. China has ordered to close the illegal mines, but this is sometimes not in the interests of local authorities, who rely on a share of the mines' revenue. However, the local country government in Fanzhi country has shut down hundreds of mines and has sent thousands of workers home.

Most of the miners are peasant farmers, who may have travelled a long way in order to make money from lucrative mining jobs. One farm worker, who was reluctant to leave, told that he could make more than 1,000 yuan ($120) a month from mining.

A source from the local transport authority told that it had arranged extra trains and buses to help more than 2,500 migrant workers daily. The local government had even sponsored the trip home for those too poor to pay. However, new illegal mines are springing up as fast as the authorities can shut them down. There are big profits to be made out of illegal mining.

A number of people wish to raise their standard of living. Increased wealth leads to additional purchasing power, which means that people collect more passions, many of which are made of metal. Finally, every year some metals are more or less permanently bound up in buildings, bridges, dams, and roadways. However, despite this continued consumption, there are several ways to extend our existing metal reserves.

CONSERVATION OF METAL RESOURCES

Various explorations have indicated that mineral deposits on the sea floor are vast. The sea floor need not be drilled or blasted, and explorations can be done under sea floor. The technology of collection is complicated and costs are expected to be high.

- ◇ Instead of metals, plastics can be used for a wide range of products such as toys, packaging materials, and furniture. Substitution, powerful as it is, is not a solution for all our problems at present. Many synthetic structural materials are more expensive than metals, which can be used only in special applications.

- ◇ An alternative approach to substitution is the reuse or recycling of materials. In the US, 400 kg of iron and steel are manufactured per person per year.

- ◇ Energy and materials are conserved if an old, worn out car engine is remachined and rebuilt rather than shredded and scrapped. Thus we can prevent the manufacture of a new automobile in its place.

- ◇ Materials resources can also be conserved if manufactured items are built to be smaller and lighter, for example, during the decade from 1975 to 1985, manufacturers reduced the overall weight of the average automobile by 25%, which led to a tremendous reduction in the consumption of a variety of resources.

Case Study 9

Artisanal and Small-Scale Mining

The MRF small-scale mining is a subsection of the Mineral Resources Forum (MRF) that focuses specifically on the issues in relation to artisanal and small-scale mining. These issues are related to environmental, health and safety, women and children's participation, organizational, technical and financial constraints, and regulatory and legal normalization.

What is small-scale mining?— This is a question which has been debated at some length but which does not have a single answer satisfactory to all interested parties. A variety of countries and organizations have defined the small-scale mining to suit their specific requirements.

Small-scale mining in the Asian region (*Source:* UN ESCAP)

The aims are to increase the awareness of the small-scale mining sector, to encourage its acceptance as a legitimate source of livelihood and to draw attention to information and events that would be useful to small-scale miners and policy makers who can make an impact on the small-scale mining sector.

Artisanal mining in South America (*Source:* UNIDO)

Case Study 10

Limestone mining in the Doon valley and Mussorie hills has left permanent scars on this famous hill, destroying forests and permanent water sources. The Rural Litigation and Entitlement Kendra in Dehra Dun filed a public interest case in the Supreme Court, and in a historic judgment, the court ordered the closure of the mines on grounds of environmental destruction.

Case Study 11

Jharia Coal Mining

Jharia coalfield situated about 260 km northwest of Calcutta in Dhanbad district of Bihar covers an area of 480 sq.km. It is the most important source of cooking coal in India, containing a reserve of approximately 13,000 million tons of coal. Located in a rocky, dry and barren zone, Jharia coalfield has been used for coal mining since 1980. In the low-lying valley, agriculture is carried out in patches but the crop yield is very poor.

The earth has perhaps 30,000 plant species with parts that people can eat. However, only 15 plants and 8 terrestrial animal species can supply an estimated 90% of our global intake of calories. The three major grain crops, namely wheat, rice, and corn provide more than half the calories that the people consume. These three grains and most other food crops are annuals, whose seeds must be replanted each year.

Two-thirds of the people survive primarily on the traditional grains (mainly rice, wheat, and corn), and not on the meat because they cannot afford. As their earnings increases, people tend to consume more grains not only directly, but also indirectly in the form of meat (mostly beef, pork, and chicken), eggs, milk, cheese, and other products of grain-eating domesticated livestock. Fish and shellfish are other important sources of food for about one billion people, mostly in Asia and in coastal areas of developing countries. But on a global scale, fish and shellfish supply only 7% of the world's food, forming less than 6% of the protein in the human diet.

FOOD AND THE GROWING POPULATION

The study reported by World Bank in 1985 identified 87 developing countries in which the proportion of the population lacking sufficient calories to have an active working life were between 10 per cent and 50 per cent with an average of 34 per cent. Sixteen per cent are at less than 80 per cent of the FAO/WHO minimum, a level known to cause stunt growth and to create serious health risks; they are so severely undernourished that they cannot lead to the normal working life, and perhaps as many as 30 per cent of the people in the world are chronically hungry. The global land use is given in the following graph (Figure 5.1). The total area of the land is 13,400 mha.

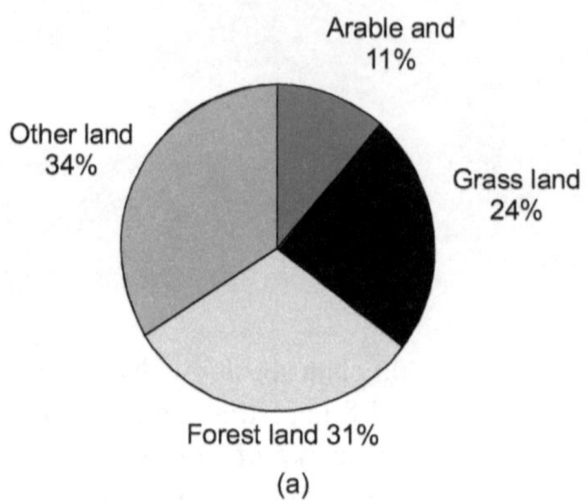

Figure 5.1   Global land use pattern

Just over a half of the world's cropland is in the developing countries that presently have three-fourths of the world's population. In any case, the food supply problems in majority of the developing countries are due to the short of land, and the potential areas for conversion are in the wrong places.

FOOD PRODUCTIVITY

Humans depend on the following three systems for their food supply:

1.   Croplands that mostly produce grains and provide about 76% of the world's food.

2.   Rangelands that produce meat, mostly from grazing livestock, and supply about 17% of the world's food

3.   Oceanic fisheries that supply about 7% of the world's food.

Since 1950, there has been a steady increase in the global food production from all these three systems. This phenomenal growth in food productivity occurred because of technological advances such as increased use of tractors and farm machinery and high-tech fishing boats and gear; inorganic chemical fertilizers; irrigation; pesticides; high-yield varieties of wheat, rice, and corn; densely populated feedlots and enclosed pens for raising cattle, pigs, and chickens; and aquaculture ponds and ocean cages for raising some types of fish and shellfish.

By 2050, to feed the world's 9.3 billion people, we must produce and equitably distribute more food than that has been produced since 10,000 years (agriculture began about 10,000 years ago), and this must be done in an environmentally sustainable manner. Some analysts believe that we can continue expanding the use of industrialized agriculture to produce the necessary food. Other analysts contend that environmental degradation, pollution, lack of water for irrigation, overgrazing by livestock, overfishing, and loss of vital ecological services may limit the future food production. A key problem is that human activities continue to take over or degrade more of the planet's net primary productivity, which supports all life.

An increase in food production is achieved by

1.   Agriculture
2.   Modern farming
3.   Green revolution
4.   Genetic engineering

Agriculture

The total food production of the earth has increased dramatically in recent years. But the population has also been increasing as well. Between 1960 and 1987, yields of wheat trippled in Europe and Asia, and the rice yields increased almost rapidly. In 1986, India imported huge quantities since starvation were rampant but today our country has a surplus. Between 1977 and 1987, agricultural production was increased to 50–60 per cent in China, representing a rate of increase far greater than the rise in population.

Despite these fortunate increases, there are still several reasons for significant alarm. One such reason is the separation between the rich and the poor, which has led to a global system where excess stock rot in some regions whereas in other regions people starved. For example, in Europe, where the population is stabilizing and agricultural production is increasing rapidly, the per capita grain production has risen sharply in the past 25 years, whereas in Africa, the population is increasing faster than the rate of food production, and today millions are starving. In addition, serious problems related to land use, soil erosion, and the availability of energy and fertilizer resources leave a little room for complacency.

From the beginning of agriculture until the early 1990s, most increase in worldwide food production have resulted into an expansion of the total area of farmland under cultivation. This rapid increase in cultivated area continued until the mid-1970s. Today there is very little room for further expansion, and global food production is facing an absolute barrier. The human population continues to grow, but nearly all of the prime agricultural land has already been brought under cultivation. In fact, in many regions, the availability of prime agricultural land is decreasing because farmlands are converted to make way for expanding cities. Thus wheat fields, orchards, and truck farms are being converted to roadways, housing developments and shopping malls.

Another problem causes great concern. Historical records show that the period between 1950 and 1985 was a time of unusually plentiful rainfall and mild climate. If the climate changes adversely, farms in many regions of the world could fail.

Sustainable land management   The most widely used method for more sustainable land management is to control the number of grazing animals and the duration of their grazing in a given area so that the carrying capacity of the area is not exceeded. However, determining the carrying capacity of a range site is difficult and costly. In addition, carrying capacity varies with factors such as climatic conditions (especially drought), still type, invasions by new plant or animal species, kinds of grazing animals, and intensity of grazing.

A more expensive and less widely used method of land management involves suppressing the growth of unwanted plants by herbicide spraying, mechanical removal, or controlled burning. A cheaper way to discourage unwanted vegetation is the controlled, short-term trampling by large numbers of livestock.

Controlling overgrazing and undergrazing   Most rangeland grasses have deep and complex root systems that help anchor the plants. These roots also extract the underground-water so that plants can withstand drought, and store nutrients so that plants can grow again after a drought or fire. Blades of these grasses grow from the base, and not from the tip. Thus, only if its upper half, called its metabolic reserve, is eaten and its lower half remains, rangeland grass is a renewable resource that can be grazed again and again. The exposed metabolic reserve of a grass plant is where the photosynthesis takes place to provide food for the deep roots of rangeland grasses. If all or most of the lower half of the plant is eaten, the plant is weakened and will soon die.

Cattle and sheep that graze on rangeland use a resource (grass) that humans cannot eat, and most of this land is not suitable for growing crops. Moreover, because of poverty, insufficient economic aid, and the nature of global economic and food distribution systems, very little, if any, additional grain grown on land used to raise livestock or livestock feed would reach the world's hungry people.

Overgrazing can limit livestock production. The feeding of large numbers of domestic livestock in a particular area for a long period causes most overgrazing. Such overgrazing lowers the net primary productivity of grassland vegetation and reduces grass cover. It also exposes the soil to erosion by water and wind, compacts the soil (which diminishes its capacity to hold water), and causes desertification.

Sometimes grasslands can suffer from undergrazing, where absence of grazing for long periods (at least 5 years) can reduce the net primary productivity of the grassland vegetation and the grass cover. Moderate grazing of such areas removes the accumulations of standing dead materials and stimulates the new biomass production.

There are two major ways to maximize the livestock productivity without overgrazing or undergrazing rangeland vegetation: 1) to control the number, types, and distribution of livestock grazing on land, and 2) to restore and improve the degraded land.

About 40% of the earth's ice-free land is rangeland that is too dry, too steeply sloped, or too infertile to grow crops. It supplies forage or vegetation for grazing (grass-eating) and browsing (shrub-eating) animals. About 42% of the world's rangeland is grazed roughly by 5 billion cattle, sheep and goats, and the remaining portion is too dry, cold, or remote from population centres to support large numbers of livestock. Livestock also graze in pastures—managed grasslands or enclosed meadows usually planted with domesticated grasses or other forage.

Livestock tend to aggregate around natural water sources and stock ponds. As a result, areas around water sources tend to be overgrazed whereas other areas will be undergrazed (Figure 5.2). The damaged lands and riparian zones can be fenced off and livestock can be moved from one grazing area to another. Ranchers can also provide supplemental feed at selected sites and locate water holes and tanks, and salt blocks in strategic places.

Figure 5.2   Rangeland overgrazing (left) and lightly grazed (right). Topsoil loss is directly associated with livestock grazing

Modern Farming

Mechanized food production is based on a technological cycle. High crop yields are needed to feed urban workers, who, in turn, can provide the technology required to maintain the high food production.

Modern farming depends on three fundamental techniques:

1.   Chemical fertilization
2.   Mechanization
3.   Irrigation

Chemical fertilization   In recent years, most of the farmers in developed countries have used greater amounts of inorganic fertilizers coupled with herbicides and pesticides. Others have felt that these practices not only are harmful to the environment but also are not economically sound. According to them, the use of herbicides and pesticides is an expensive and inefficient way to control pests, and the loss of humus due to the use of chemical fertilizers ultimately reduces the soil fertility and productivity.

An organic farmer is the one who uses only organic fertilizers such as manure, bone meal, and waste plant products and applies no synthetic pesticides or herbicides.

Mechanization   High agricultural yields using specific varieties of seed has been the result of an active fertilizer program and pest control. Nearly all the commercial farmers in developed countries rely heavily on machinery. Because the use of machinery saves money in the regions where fuel and equipment are relatively inexpensive. In addition, farmers can cultivate or harvest large tracts of land quickly using tractors, and thereby take advantage of short periods of favourable weather. If farmers were to revert to labour-intensive practices, the prices of foods would rise significantly.

Irrigation   Irrigation is almost as old as agriculture itself. Today a large portion of the world's crops of vegetables and some grains depends on irrigation.

Green Revolution

As various farmers replanted only the seeds of the largest and the healthiest plants, modern strains of wheat, rice and other grains were developed gradually over the centuries. These grains were generally well adapted to local growing conditions. They were genetically adjusted to peculiarities in soil temperatures, and they were at least partially resistant to local diseases and pest infestations. However, as population increased in the nineteenth and twentieth centuries, it became apparent that traditional farming practices were not adequate to feed the world's people. Therefore, agriculturists searched for new ways to increase crop yields.

In the mid-1960s, a team of scientists in Mexico and the Philippines developed new varieties of wheat and rice that were adaptable to tropical climates and were capable of producing higher yields than any native grains. These varieties have short, upright leaves, and so plants can be grown close together without shading each other. In addition, their stalks are short and thick so that they will not bend and break when they are heavily fertilized. The potential yields of these new varieties are so spectacular that people have heralded their introduction as the **Green Revolution**.

To understand the total impact of the Green Revolution, it is necessary to understand that a **wonder seed**, by itself, does not produce large quantities of food. The seed carries only the genetic potential for high-grade production. In order to produce large quantities of food, the plant must be fertilized, heavily watered, and protected from disease and insects.

If a farmer has the money to buy new vigorous seeds as well as the necessary fuel, fertilizers, pesticides and irrigation facilities, and if he or she also knows how to manage a modern integrated farming program, then high yields and an eventual profit can be realized; otherwise, the green revolution would be of no help.

Farmers can produce more food by farming more land or by getting higher yields per unit of area from the existing cropland. Since 1950, most of the increase in global food production has come from increased yields per unit area of cropland in a process called the Green Revolution. The increase in food production by Green Revolution involves three steps.

1.   Develop and plant monocultures of selectively bred or genetically engineered high-yield varieties of key crops such as rice, wheat and corn.

2.   Produce high yields of these crops by using large inputs of fertilizers, pesticides, and water.

3.   Increase the number of crops grown per year on a plot of land through multiple cropping.

This high-input approach dramatically increased the crop yields in most of the developed countries between 1950 and 1970, i.e., during the first Green Revolution. A second Green Revolution has been taking place since 1967 by introducing fast-growing dwarf varieties of rice and wheat, specially bred for tropical and sub-tropical climates, in several developing countries. With enough fertile soil and fertilizers, water, and pesticides, yields of these new plants can be two to five times those of traditional wheat and rice varieties. The fast growth also allows the farmers to grow two or even three crops per year (multiple cropping) on the same land. Producing more foods on less land is also an important way to protect biodiversity by saving large areas of forests, grasslands, wetlands, and easily eroded mountain terrain from being used to grow food.

The increase in yield depends not only on fertile soil and ample water but also on high inputs of fossil fuels to run machinery, to produce and apply inorganic fertilizers and pesticides, and to pump water for irrigation.

Hence, high-input green revolution agriculture uses about 8% of the world's oil output.

The Green Revolution has clearly increased the food production throughout the world. Without these grains, hunger and starvation would be much more severe than they are. However, the green revolution has its problems. Perhaps the most serious concern is that recent gains in agricultural productivity have been possible only with the massive use of fossil fuels.

Genetic Engineering

In recent years, scientists are working to create new green revolutions—actually gene revolutions—by using genetic engineering and other forms of biotechnology to develop genetically improved strains of crops. Genetic engineering or gene splicing of food crops is the insertion of an alien gene into a commercially valuable plant or animal to give it new beneficial genetic traits. Such organisms are called genetically modified organisms (GMOs), e.g. pest-resistant plants, insect-resistant plants, drought-resistant plants, etc.

Staple food crops such as corn, wheat, and rice could be engineered to manufacture their own nitrogen fertilizer. Perhaps the yield could be increased through improvements in the way in which the plants use the solar energy during photosynthesis. It is even possible to imagine a drought-resistant, nitrogen-producing plant that has the roots of a potato plant, the leaves of spinach, and the fruit of a tomato.

Compared to traditional cross-breeding, gene splicing takes about half as much time to develop a new crop or animal variety, cuts costs, and allows the insertion of genes from almost any other organism into crop or animal cells. Scientists are also using advanced tissue culture techniques to produce only the desired parts of a plant such as its oils or fruits.

Ready or not, the world is entering the age of genetic engineering. The United States is the world's largest producer of genetically modified crops. Nearly two-thirds of the food products on US supermarket shelves contain genetically engineered crops, and the proportion is increasing rapidly.

However, there is growing controversy over the use of genetically modified foods (GMF). They are seen as a potentially sustainable way to solve world food problems by their producers and investors and are called potentially dangerous "Frankenfood" by the critics.

PROBLEMS IN FOOD PRODUCTION

Effects of Fertilizers

Fertilizers are the chemical nutrients added to soil to substitute loss of nutrients in course of time of agriculture. There is still a lot of misunderstanding and confusion about mineral fertilizers. In the Netherlands, it was amazed to hear several students saying that raising the crop yields with fertilizers was very dangerous and even immoral, particularly for African soils. It is time to dispel some myths about mineral fertilizers, to appreciate the role they play in feeding the world, and to assess how best they can help agriculture in meeting the challenges it faces in the decades ahead.

The world population will probably peak at some 8,000 million around 2030, when two out of every three people will live in towns and cities. Rising incomes will create a higher demand for food, meaning that over the next three decades, food production will need to increase by about 60%. Nearly all of the increase in production will have to come from developing countries through intensification of agriculture, i.e., more yield per unit time and per unit area. As urbanization reduces the rural workforce, agriculture will also need to adopt new forms of mechanization and shift to land use intensification, with all of its connotations. Those scenarios point to an increase in the use efficiencies of all natural resources, particularly water, and to the need for greater—although not proportionally greater—-use of mineral fertilizers.

Half a century ago, farmers applied only 17 million tons of mineral fertilizers to their land. Today, they apply eight times as much. In northern Europe, the use of fertilizers has increased the yield of crops from about 45 kg/ha to 250kg/ha since 1950. Simultaneously, in France the yields of wheat have increased every year, from about 1.8 tons/ha to more than 7 tons/ha. The growth in fertilizer use is certainly lower than the increase in yields. This confirms the overall pattern of increasing efficiency in using fertilizers.

Fertilizer application currently accounts for 43% of the nutrients that global crop production extracts each year, and the contribution may be as high as 84% in the years to come. Contrary to some public opinion, non-mineral nutrient sources are not likely to challenge mineral fertilizers in the future.

Though the availability of non-mineral sources such as manure and organic wastes will be more with an increased livestock production and urbanization respectively, their efficiencies are considerably lower and the current cost of using wastes for crops is still quite high.

Organic agriculture, which eliminates the use of synthetic inputs, does not appear to be a feasible alternative. A large amount of land would have to be brought under rotation with legumes or under animal production to make up for the lack of mineral fertilizer. While organic agriculture does fill a niche market, its limits and its dangers, in terms of nutrient depletion need thorough review.

At the World Food Summit (WFS) in 1996, governments committed themselves to halving the number of hungry people by the year 2015. There is a direct link between that WFS goal and fertilizer applications. Possibly, it means an 8% increase in fertilizer applications compared to the "business-as-usual" scenario. That does not seem very much, but in terms of its tonnage, it is considerable. Enhanced fertilizer use to meet WFS goals is particularly important in countries such as China and India, which make up a large proportion of the world population. But it may be even more important in Africa, where increase of 2.7% or more a year are needed in order to make up for nutrient losses, and in the humid tropics, where unfertilized annual cropping takes a heavy toll of soil organic matter.

Integrated management of production systems offers a proven path to greater fertilizer use efficiency. Farmers are being trained to observe the real impact of nutrient application rather than, for example, applying more and more urea simply because it is the cheapest fertilizer. Farmers also need to understand the effects of over-use of nitrogen on certain pathogens and other stress factors in crops. This may convince them of the need to buy non-nitrogen fertilizers and to adopt much more balanced fertilizer applications.

Effects of Pesticides

The term pesticides generally refer to the insecticides, herbicides, rodenticides, and fungicides, used to eradicate bugs, weeds, rats and voles, and fungus that plague and cut into the farming profits. Pesticides (chemicals designed to kill insects) were produced specifically to alter or kill living organisms.

In 1985, nearly 500,000 tonnes of insecticides and fungicides were sprayed on to the farmlands in the United States alone. Although the sprays permitted today are generally less persistent and more specific than DDT and other products that are now banned, most are still toxic to many different non-target species. In addition, many of the pesticides have found their way into drinking water supplies. Most other countries permit the use of wide variety of pesticides that are banned in the United States, including DDT and related products.

Scientists are certain that high concentrations of most pesticides are acutely toxic to humans. Symptoms of pesticide poisoning include loss of muscle control, nervous disorders, loss of mental awareness, and eventual death. Experiments have shown that various pesticides induce cancer in laboratory animals.

The National Academy of Sciences, USA, has concluded in 1987 that pesticide contamination of food supply may be responsible for up to 20,000 cancer cases each year, birth defects, mutations, or damage to the kidneys, liver, nervous system, or the immune systems.

Pesticides are also seen in places other than on the food we eat. The most widely used herbicides such as atrazine and alachlor are known to be potent carcinogens. Pesticides that miss their target pests can end up in air, surface water, groundwater, bottom sediments, food, and non-target organisms, including humans and wildlife. There is a great risk of contamination of groundwater. The risk is particularly acute in rural areas, where millions of people get their drinking water from private wells that are rarely treated or monitored.

Insects have evolved mechanisms of detoxifying and resisting the pesticides. Insect pests are also controlled by a wide variety of predatorsand disease (biological control), natural controls are not always effective, and severe crop losses have occurred throughout the history. Therefore scientists have devised alternative control measures. The modern pesticide era started in about 1940, when a chemical called DDT (Dichloro diphenyl trichloroethane) was found to be a potent insecticide. DDT was used to eradicate malaria in India and Pakistan and Black disease in Africa. It has been estimated that all the agricultural products sold in Punjab has considerable amount of DDT.

Case Study 12

Given here are some of the food poisoning cases. Many fertilizers and pesticides have threatened human health because they may be carried through the food chain and may cause food poisoning.

Microbe-contaminated food and water kill up to two million children in developing countries each year

In March 2003, 172 children were reported with food poisoning in southern Vietnam after eating their lunch in their school cafeteria.

The same week, 110 people in St. Louis, USA, fell ill during a Mardi Gras dinner, and 17 labourers in Bombay suffered crippling stomach pains from eating tainted sweets.

In the US, health authorities requested the recall of animal feed supplements containing "elevated levels" of dioxin (the toxic chemical apparently leaked from industrial waste dumps), and Italian police seized the buffalo milk contaminated with dioxin. High levels of pesticides were found in samples of spinach taken from supermarket shelves in the UK, and in bottled water in India.

In Europe, the "mad cow" disease emergency and recent alarms over dioxin contamination in meat have exposed the vulnerability of the primary food production. In the US, food poisoning outbreaks traced to ground beef contaminated by mutant *Escherichia* bacteria in cattle faeces has raised public concern over slaughterhouse procedures. "The intensification of plant, livestock and fishery production may also increase the risk of chemical contamination with pesticides and veterinary residues."

And the families of two victims of the human variant of "mad cow" disease announced legal action against a steakhouse chain accused of serving banned beef.

Biological control

Natural Enemies   Nowadays, the technique of encouraging predator populations to control insect pests is generating renewed enthusiasm because of the effects of sprays and of the resistance developed by the pests.

Sterilization   The pests can be controlled without being killed directly if the adults of one generation are sterilized.

Hormones   Many insects begin their life in some larval stages and later change (metamorphose) into mature adults. For example, a caterpillar is a larva that later matures to become a moth or butterfly. The changing stages in an insect's life are controlled by chemicals called hormones. If insects are artificially sprayed with specific hormones, their life cycles can be disturbed, and eventually such use of spray can be lethal. Therefore, hormones can be used as insecticides.

Effects of Waterlogging

Waterlogging is a major problem found with irrigation. Farmers often apply large amounts of irrigation water to leach the salts deeper into the soil. Without adequate drainage, however, water accumulates underground and gradually raises the water table. The saline water then envelops the deep roots of plants thereby lowering their productivity and killing them after prolonged exposure. At least one-tenth of the irrigated land worldwide suffers from waterlogging, and the problem is getting worse.

Excessive use of canal irrigation has disturbed the water balance and has created waterlogging problem as a result of seepage or rise in the water table. From this, it is clear that waterlogging may be due to surface flooding or due to high water table. In waterlogged soils, the oxygen available for respiration of plants is less.

Effects of salinity   Approximately, 17% of the world's cropland that is irrigated produces almost 40% of the world's food. The irrigated land can produce crop yields two to three times greater than those from rainwater. But irrigation also has a downside.

The majority of irrigation water is a dilute solution of various salts picked up, as the water flows over or through soil and rocks. These salts in smaller quantities are essential nutrients for plants, but are toxic in large amounts.

In arid areas with low rainfall, poor drainage and high temperatures, the unabsorbed irrigation water evaporates quickly, leaving behind a thin crust of dissolved salts like sodium chloride in the topsoil. This accumulation of salts is called salinization (Figure 5.4). In other words, salination is the increase in the concentration of soluble salts in the soil.

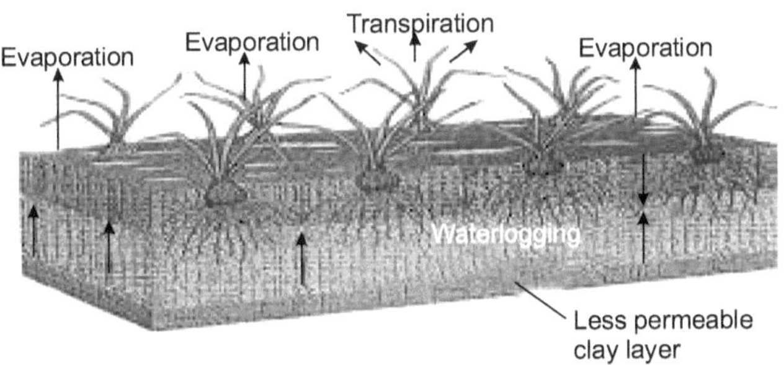

Figure 5.4   Salinization due to waterlogging

Excess of these salts form a white crust on the soil surface and are injurious to the survival of plants. The water absorption process of the plant is severely affected. It directly affects the productivity of soils by making the soil unfavourable for good crop growth. Indirectly, it lowers the productivity through adverse effects on the availability of nutrients and on the beneficial activities of soil microflora.

The loss in crop production due to salinity in India amounts to 6.2 million tons (FAO data) and 9.7 million tonnes (Indian data). In India, the places which are frequently waterlogging are the coastal areas of Mumbai and Kerala, estuarine deltas of Ganges, and Andaman and Nicobar Islands. According to a 1995 study, severe salinization has reduced the yields on 21% of the world's irrigated cropland, and another 30% has been moderately salinized. The most severe salinization occurs in Asia, especially in China, India, and Pakistan. In the United States, salinization affects 23% of all irrigated cropland. However, the proportion is much higher in some heavily irrigated western States. For example, salinization affects 66% of the irrigated land in the lower Colorado Basin and 35% of such land in California.

Though sufficient water is present in the soil, it is pathologically unavailable to the plants because of the higher concentrations of soil solution. In India, about 6–7 million hectares of saline land is present. Due to intensive agricultural management practices, thousands of hectares of good agricultural land turn saline every year in Punjab alone. The control measures used to prevent the salinity are the following:

◇   Knowing the ways to prevent and deal with soil salinization.

◇   Reducing irrigation.

◇   Switching to salt-tolerant crops such as cotton, sugarbeat, and barley.

◇   Flushing soil which is expensive.

◇   Not growing crops for 2 to 5 years.

◇   Installing underground drainage system.

Moreover the salinated lands can be reclaimed by the simple process of leaching with plenty of water, i.e., heavy irrigation, which leaches the salts at the surface to greater depths.

ECOSYSTEM HEALTH

The occurrence of significant forms of disease or deformities (Figure 5.4) among wildlife can be an indication of deteriorating ecosystem health. The disease emergence is an early warning of challenges to the integrity of the biological systems that comprise ecosystem sustainability and that provide for economic growth and healthy human populations.

Various estimates suggest that between 900,000 and 2 million people become ill each year in the United States by ingesting the protozoan, bacterial, and viral pathogens in an incompletely treated or untreated drinking water.

Figure 5.4   A mink frog from Minnesota with an extra limb. This is one of several types of deformities being found in a high percentage of amphibians at various North American locations

Food Hazards

Food-borne diseases remain responsible for high levels of morbidity and mortality in the general population, but particularly for at-risk groups, such as infants and young children, the elderly and the immunocompromised. Some examples of causative agents that cause serious food hazards are discussed below.

Zoonotic agents

◇   *Salmonella*   It is a microbe usually transmitted in foods contaminated with animal faeces, and causes nausea, vomiting, and diarrhoea.

◇   *Prion proteins*   The infective agent in beef believed to cause the Creutzfeldt–Jakob variant of "mad cow" disease in humans

Food-borne pathogens

◇   *Campylobacter jejuni*   The most common pathogen in an uncooked chicken causing arthritis and a rare disease of the nervous system. 

◇   *Listeria monocytogenes*   It contaminates the raw meat and vegetables, and affects mainly the pregnant women, the newborn and the elderly.

Emerging pathogens

◇   *Salmonella enteritidis*   It infects the eggs in hens' ovaries and causes fever, abdominal cramps, and diarrhoea.

◇   *Escherichia coli O157:H7*   It is found in the intestines of cattle, normally gets transmitted from ground beef, and can cause permanent kidney damage and death.

Industrial contaminants

◇   *Dioxin*   The toxic chemical formed from combustion of municipal wastes and fuel, and transmitted through intake of animal fats. It causes skin disease, liver damage and possibly cancer.

Agricultural chemicals

◇   *Pesticides*   They may suppress the immune responses to an infection, making people more vulnerable to the disease.

◇   *Veterinary drugs*   These are the residues found in food, e.g. antibacterials, steroids, hormonal growth promoters and beta-agonists.

Allergens

◇   In the US, many children and adolescents are dying from the anaphylaxis induced by peanuts and nuts than from the insect stings.

FOOD SAFETY

When an unsafe or contaminated food enters the food chain, it is distributed more rapidly and a greater number of consumers are exposed to risk. The first link is human. Complementing the food chain approach, FAO is developing a set of **Good Agricultural Practices** (GAP) that establish basic principles and indicators for on-farm operations such as soil and water management, crop and animal production, storage, processing, and waste disposal. While GAP has multiple objectives including sustainable use of natural resources and contributing to rural livelihoods, it could contribute significantly to food safety and quality. Thus, good agricultural practices apply the available knowledge to utilize the natural resource bases in a sustainable way for the production of healthy food and non-food products.

The Food Safety Department strives to reduce the serious negative impact of food-borne diseases worldwide. Food and water-borne diarrhoeal diseases are the leading causes of illness and death in less-developed countries, killing approximately 1.8 million people annually, in which most of them are children. In order to reduce the incidence and economic consequences of food-borne diseases, the WHO Department of Food Safety has been assisting the Member States to establish and strengthen their programmes for assuring the safety of food from initial production to final consumption.

Recent trends in global food production, processing, distribution and preparation are creating an increasing demand for food safety research in order to ensure a safer global food supply. Food Safety Department works with other WHO departments, Regional Offices and WHO collaborating centres as well as other international and national agencies. In particular, the WHO works closely with the Food and Agriculture Organization of the United Nations (FAO) to address food safety issues along the entire food production chain—from production to consumption—using new methods of risk analysis. These methods provide efficient, science-based tools to improve the food safety, thereby benefiting both the public health and economic development.

Five Keys to Safer Food

The good hygiene foods can prevent the transmission of pathogens responsible for many food-borne diseases. Governments, industries and consumers have shared the responsibility in ensuring the safety of food. The WHO has long been aware of the need to educate all food handlers, including the professionals and ordinary consumers, about their responsibility in food safety. After nearly for a year of consultations with food safety experts and risk communicators, the WHO introduced the Five Keys to Safer Food in 2001. The WHO Five Keys to Safer Food are the simple rules elaborated to promote safer food handling and preparation practices.

1. Keep clean.
2. Separate the raw foods from the cooked foods.
3. Cook thoroughly.
4. Keep the food at safe temperatures.
5. Use clean and safe water and raw materials.

However, this simple message is not enough and in October 2004, the WHO produced a draft version of a basic training manual for the professionals, teachers and other interested organizations to train the food handlers and consumers, including the school children. The 5-keys manual, entitled "Bring Food Safety Home", has two objectives:

1. To provide a generic food safety training material that can be used as a framework to produce the food safety training materials for a variety of audiences at the national level.

2. To provide recommendations on how this basic material can be adapted for different audiences based on the social, economic and cultural differences between the countries.

Ecologically man is only a part of energy flow in nature. Man requires energy for his daily needs. Energy is the primary input for almost all economic activities and is, therefore, vital for improvement in the quality of life. During the early stages of human civilization, the daily per capita need for energy was less. During the agricultural stage, the muscular energy of domestic animals was used for work. The per capita energy consumption gradually increased. In the nineteenth century, i.e., during the industrial stage of human civilization, the use of fossil fuels started and the per capita energy requirement increased to about 70,000 kilo calories per day.

The energy demand is not an exception to the economic theory of limited means and unlimited wants. The pace of exploitation of the energy resources has been growing over time, and has resulted in gradual depletion of the scarce reserves. The critical link between energy and economy has exposed the vulnerability of the nations to the volatile energy situation. Today, energy has become a key factor in deciding the product cost at the micro-level as well as indicating the inflation and the debt burden at the macro-level. The cost of energy is a significant factor in the economic activity, at par with factors of production like capital, land and labour. In an energy shortage situation, the authorities call for the energy conservation measure, which essentially mean using less energy for the same level of activity.

GROWING ENERGY NEEDS

Global Scenario

Approximately 80% of the world's energy is produced by the fossil fuels or conventional resources. However, in France, the nuclear reactors are installed which produce enough energy to meet 70% of the country's requirements. The World's demand for oil rose from 436 million tons in 1960 to 3200 million tons in 1999. The corresponding figures for coal are 1043, 2146 and for natural gas are 187, and 2301. The demand will continue to grow. Of the developing countries, China has the highest per capita consumption of energy. For India, per capita consumption is lower than that of China.

The non-conventional resources include solar (2.6%), wind (67.4%), biomass (16%) and small hydro-sources (13.1 %) and energy from waste (0.9%). Among the non-conventional resources, hydropower is the largest. Hydropower projects are in operation both in the developed and in the developing countries; notable among the latter are China, India and Brazil. The Hydropower potential is huge, and at present only 15 per cent of the potential in the developing world is being utilized.

Wind power has also a great potential. Windmills and sails have been in use since ancient times. It is a fast-growing resource. If the present trend continues, wind power could supply 10% of the world's electricity, by 2020.

The use of solar energy is through photovoltaic cells. *The Photovoltaic News* reported that the world's photovoltaic production increased from 0.1 MW to 200 MW in 1999.

The biomass resources constitute various types of cultivated or uncultivated vegetations. Biomass includes large quantities of cattle dung and human excreta, agricultural residues, biogases and other by-products of agro-based industries. Wood forms the chief resource and is the primary fuel for the people in Africa and Asia. Excessive use of wood has led to the depletion of forests.

Indian Scenario

Coal, oil, gas and water constitute the main sources of energy in our country. The share of various energy sources in the commercial consumption of energy is mostly from coal (56%) and petroleum (32%), the other sources being nuclear, natural gas and water. Apart from commercial energy, a large amount of traditional energy sources in the form of fuel woods, agriculture wastes and animal residues are used. The share of energy resources for the urban and rural areas in our country in the 1980s was in the ratio of 1 : 1, but with rapid industrialization. When growth and continuous gap in demand and supply the commercial sector has surpassed its counterpart and in the year 2008-09 the ratio is estimated to be around 4 : 1.

NON-RENEWABLE ENERGY SOURCES

Coal, oil, gas and hydroelectric potential constitute the conventional (non-renewable) resources for electricity generation in the country. Among them, power plants based on coal are adopted in some areas. Similarly, hydropower is the main source for the electricity generation. Also, oil, natural gas and nuclear power partly contribute to the resources.

Coal

The coal resources are assessed to a tune of 186 billion tons in India. Apart from that, about 5 billion tons of lignite deposits are available for power generation. With this, the power requirement for the next 100 years could be able to met out. According to estimates, although the coal is abundant in nature, it will last only 200 years. It is low in calorific value and its shipping is expensive. It is a pollutant and when burnt it produces CO_2 and CO. Extensive use of it as a source of energy is likely to disturb the ecological balance, as the vegetations are not capable of absorbing such large proportions of carbon dioxide produced by burning large quantities of coal.

Oil and Natural Gas

India has large oil and gas bearing sedimentary basins to a tune of 728 million tones of crude oil and 686 billion cubic metres of gas. Private sectors have entered into the field that expects additional resources. The existing oil companies in Arab countries are keen in modernizing the technology, and in adopting new strategies for the conveyance of natural gas through pipelines. Oil, the golden resource (also called the liquid gold), is exhaustible, i.e., a limited treasure of humankind is depleting. It has become evident that the oil reserves of the world are running out at a rapid pace. Our country, in view of the ever-increasing oil imports bill, is realizing that increasing dependence on oil has to be checked, and every drop of petroleum has to be put to its optimum use. The consumption of petroleum products (Table 6.1) has increased sharply from 22 million tons in 1975–76 to approximately 90 million tones in 2005. Because of the rising demand of petroleum, the domestic crude oil production could not keep pace with it.

Table 6.1   Consumption of petroleum products

Transport	40%
Industries	25%
Household	18.6%
Agriculture	8.7%
Others	7.7%

The hydel power generation is assessed to a magnitude of 60 per cent by about 84,000 kW annually. However, the power generation is limiting to a tune of about 15 per cent only, and 7 per cent are under execution, and the rest are to be exploited.

Electricity is one of the most commonly used forms of energy. Since independence, India has increased the capacity of electricity generation to over 91,200 MW in 1994–95, out of which thermal constitutes 70%, hydel 27% and nuclear 3%. Despite the massive growth in power generation, the country is still facing a power shortage.

Table 6.2 illustrates where India stands in terms of installed capacities and per capita consumption against some major countries of the world.

Table 6.2   Installed capacities and per capita consumption in different countries

Country	Installed Capacity (MW) (as of 1991)	Per Capita Consumption (kWhr) (as of 1991)
India	60,060	243
Japan	179,600	6,314
Australia	15,170	5,761
France	104,100	6,096
Germany	118,120	5,914
Sweden	32,700	14,105
UK	70,010	5,059
Canada	98,470	15,472
USA	693,020	10,608

EXPLOITATION OF FOSSIL FUELS

Today the world's energy resources have reached critical stage. Most of the world's human population uses fossil fuels (coal, petroleum and natural gas). The fossil fuel resources are being rapidly depleted. As a result, these resources may last only for another few centuries. This has led to the search of alternate sources of energy. The commercial energy consumption has increased to 30 per cent from that of 1998. The main drivers of this increase are the accompanying structural changes of economic growth and a rise in population together with rapid urbanization. Its use in sectors such as industry, commerce, transport, telecommunications, and a wide range of agricultural and household services has compelled us to focus our attention to ensure its continuous supply to meet our ever-increasing demands.

Industrial Sectors

The largest consumer of energy, consuming about 50% of the total commercial energy produced in the country, followed by the transport sector. The most energy-intensive industries which together account for nearly 80% of the total industrial energy consumption are of the fertilizer, aluminum, textiles, cement, iron and steel, pulp and paper and chloro-alkali.

Transport Sector

It is the largest consumer of petroleum products, mainly in the form of high-speed diesel and gasoline, and accounts for nearly 50% of the total consumption.

Agricultural Sector

With increased mechanization and modernization, the agricultural sector's consumption of commercial energy has grown considerably. The share of the farm sector in electrical energy consumption has increased from a mere 3.9% in 1950–51 to about 32.5% in 1996–97.

The Domestic Sector

In the domestic sector, the consumption of natural fuel (mostly wood) energy is very high. Around 78% of rural and 30% of urban households depend on firewood. However, the mix of traditional fuels in the national energy mix is decreasing, as some more efficient commercial fuels are increasingly substituting these. In particular, between 1970–71 and 1994–95, the annual consumption of electricity per household went up from 7 kWh to 53 kWh; of kerosene from 6.6 kg to 9.9 kg; and of cooking gas from 0.33 kg to 3.8 kg. There is, however, a marked disparity in the level of energy and the type of fuel consumed in rural and urban areas.The percentage use of various sources for the total energy consumption in the world in given in the following Table 6.3.

Table 6.3   Percentage use of various energy sources

Sources	Percentage of total energy consumption
Coal	32.5
Oil	38.3
Gas	19.0
Uranium	0.13
Hydro	2.0
Wood	6.6
Dung	1.2
Waste	0.3

On one hand, the demand for energy is increasing, whereas on the other hand, the energy resources are becoming scarce and costlier. This steady increase in gap has compelled the technocrats and decision-makers in the industry not only to develop new measures of energy conservation but also to have a systematic approach towards the present trend of energy consumption through energy auditing, and application of modern techniques and methods for minimizing the energy wastage.

FUTURE ENERGY SITUATION

The energy situation in the near future would be characterized by the following:

◇ Supply of oil will fail to meet the growing demand before the year 2010.

◇ The demand for energy will continue to grow. This needs to be satisfied by energy resources other than oils, which will be progressively conserved for the uses that only the oil can satisfy.

◇ The continued growth of energy demand requires new energy resources other than the oil. Hence, they must be developed with the utmost vigour. The change from a world economy dominated by oil must start now. The alternatives will require 10 to 15 years to develop, and the need for these replacement fuels would increase rapidly in the following decades.

◇ Electricity from nuclear power is capable of making an important contribution to the global energy supply although worldwide acceptance of it on a sufficiently large scale has yet to be established.

◇ Coal has the potential to contribute substantially to the future energy supplies. Coal resources are abundant, but taking advantage of them requires an active programme of development by both the producers and consumers.

◇ Other than hydroelectric power, renewable resources of energy such as solar and wind power are unlikely to contribute significant quantities of additional energy during this century at the global level, although they would be important in specific areas. They are, however, likely to become increasingly important in the twenty-second century.

◇ Energy efficiency improvements can further reduce the demand and narrow the prospective gaps between the supply and demand. Policies for achieving energy conservation should be the key element of all future energy strategies.

On the energy front, despite a 50 per cent increase in global demand, energy resources will be sufficient. But there will be major changes in the geopolitics of energy.

Asia (especially China and to a lesser extent, India) will drive the expansion in energy demand, replacing North America as the leading energy consumption region and accounting for more than half of the world's total increase in demand.

- By 2015, only one-tenth of Persian Gulf oil will be directed to Western markets whereas three-quarters will go to Asia.

- The United States and other Western countries will increasingly rely on Atlantic Basin sources of oil.

RENEWABLE SOURCES OF ENERGY

Since there is a fast depletion of conventional sources of energy, some alternative sources of fuel are used. They are generally called renewable/non-conventional sources of energy. They include solar energy, hydroelectric energy, geo-thermal energy, wind power, tidal energy, energy from biomass, dung energy, and nuclear energy. Solar energy is a powerful alternative to fossil and nuclear fuels.

The country possesses large quantity of renewable energy resources. It is estimated that about 1,26,000 MW power potential are anticipated through wind-based generation, which has begun to be exploited in southern and western regions is estimated at 20,000 MW, small hydro-potential at 10,000 MW and biomass energy at 17,000 MW. The largest source, however, is of ocean, thermal, tidal and wave power technologies, which are yet to be established through pilot schemes. Subject to cost-effective technologies being developed, solar energy is also available in abundance for tapping.

Solar Energy

Sun is an inexhaustible and pollution-free source of energy. Solar equipment have been developed to harness the sun's rays to heat water, cook meals, light our houses and run certain machines. Solar energy has the greatest potential of all the sources of renewable energy, and if only a small amount of the form of energy could be used, it will be one of the most important supplies of energy. Its potential is 178 million MW, i.e., about 20,000 times the world's demand. Utilization of solar energy is of great importance to India since it lies in a temperate climate of the region of the world where sunlight is abundant for a major part of the year.

Ever since the photoelectric effect was discovered, the conversion of light into electricity is known. Its energy can be utilized as thermal and photovoltaics. Solar photovoltaic cells can be used for conversion of solar energy directly into electricity. When the solar insulation is high, and in remote areas where the establishment of the conventional power line is expensive, the photoelectric conversion may prove to be advantageous. In industrial lighting, solar cell lighting will save the expenses of laying cables and transmission, and with the future projected costs, solar lighting may be economical. In deserts and in remote areas, solar cell would be the economic proposal. At present railway signal lighting and rural and industrial outside lighting involve the use of solar cells. Rural communication, unmanned TV transmitters and refrigeration in remote areas are some of the applications underway.

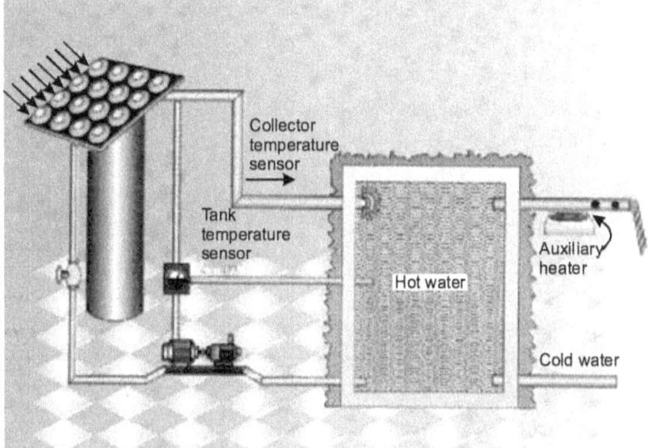

Figure 6.1   Solar water heater

Solar direct heating is yet another application. Though not very common, the solar heating and cooling of buildings with the help of solar heating and cooling systems are accepted. Solar water heaters (Figure 6.1) are in common use in commercial and domestic sectors. Solar cookers have also been developed and find increasing applications.

Nowadays, solar energy has numerous applications that are as follows.

⋄   Solar water harvesting

⋄   Solar heating of building

⋄   Solar distillation

- ◇ Solar pumping
- ◇ Solar drying in agricultural and animal products
- ◇ Solar furnaces
- ◇ Solar cooking
- ◇ Solar electric power generation
- ◇ Solar thermal power production
- ◇ Solar green houses.

Case Study 13

Solar Energy in India

India is undergoing a major social and economic revolution because of indigenously developed solar energy and biomass conversion technologies. A solar cooker consisting of an aluminum reflector (10 sq. feet in area) has been tried in India.

- ✶ The solar passive architecture and building introduced by the Development of Non-conventional Energy Sources (DNES) in the extremely cold and remote areas of Ladakh, have opened entirely new vistas. They provide comfortable housing and fresh vegetables to the paramilitary and military personnel serving in the forward areas as well as to the civilians residing in the area.

- ✶ Further, the breakthrough achieved by Indian science and technology in tapping solar and wind energy on the ice continent of Antarctica, may well provide the solution for taking the boon of power to cold, remote, desolate and inaccessible areas like Leh in Ladakh and outposts in the Himalayan range.

- ✶ Rural electrification through solar photovoltalics continued to be emerged as a strong option for remote and far-flung areas. More than 320 villages in Rajasthan, U.P., Gujarat and Haryana are being supplied with street-lighting units now.

- ✶ India's first solar power station to be in operation soon near Delhi is a great breakthrough in indigenous technologies of non-conventional use of powers.

- ✶ Favourable results have been received in using solar energy for desalination to supply water to distant cattle rearing station in the desert of Central Asia and Kazakhastan.

- Solar energy can also be used for manufacturing microwater plants that are used as a supplementary protein for feeding cattle.

- The Department of Metallurgy of the College of Engineering in Pune designed a solar furnace that can generate heat up to 2000°C. This is claimed to be the first of its kind in India.

- The first solar water heating system was installed in 1982 at the Haryana Breweries Limited at Murtha. The system is capable of heating 15,000 litres of water per day at an average temperature of 65°C. Another system capable of heating 25,000 litres of water at 65°C was installed at the Medical College, Rohtak.

- The first solar desalination plant in the state is under installation at the Haryana Tourist Complex at Dubchick. The plant will treat saline water and will have capacity of 2,000 litres of potable water per day.

- The first multipurpose solar dryer cum warehouse in the state is currently under construction at Ganaur, which can be used to dry the chilies, potato chips and other foodstuffs.

- A photovoltaic irrigation pumpset has been installed at the Haryana Agricultural University.

Case Study 14

Ethanol from "MADHURA" Sweet Sorghum

A solar-powered ethanol distillation plant was set up in the Nimkar Agricultural Research Institute (NARI) campus.

Biomass gasifier

Large amount of agricultural residues like sugarcane leaves, Bajra residues, wheat residues, etc. are burnt in the fields. Besides creating the air pollution, there is a tremendous loss of energy. Nimkar agricultural research institute (NARI) has therefore developed a biomass gasifier for loose agricultural residues.

Attributes of NARI gasifier

- It is a thermal gasifier of 800 kW capacity.
- It is a multi-fuel gasifier and can run on sugar cane leaves, bajra husk, safflower residues, sweet sorghum stalks and bagasse, sugar cane bagasse, etc.

- About 20–24% of the fuel is converted into char, which is a value added item.
- Zero waste water system. Hot gas cleaning.
- PLC-controlled unit. Only two operators per shift are required.

Case Study 15

Solar Detoxification of Distillery Waste

The distillery wastes, which appear black, smell very obnoxious and have high chemical oxygen demand (COD), are cleaned using chemicals and solar energy. A pilot scale plant has been set up which can clean 100–200 litres of diluted distillery wastes in two days. The chemical can be recycled and the waste-water has 85–90% transmittance (depending on solar energy), and hence, is fit for discharge in waterways. Initial experiments with the pulp paper effluent have also yielded excellent results.

Nuclear Energy

Controlled atomic fission and fusion is the basic principle used in the atomic power generation. The fusion of the atoms of certain elements such as Uranium-235 releases enormous amount of energy. The enormous heat produced during the reaction could be used to generate vapour of the working fluid, which will run a turbine coupled to a suitable electric generator. Similarly, the fission of 1 amu (atomic mass units) Uranium-235 can generate energy which is equivalent to that obtainable from burning of 15 metric tons of coals or about 14 barrels of crude oil. Fission reactors are popularly employed in various countries. Table 6.4 lists the percentage of nuclear energy generated in different countries. Though nuclear energy constitutes less than 5% of our country's energy requirements, plans are under various stages of execution to have the nuclear generation to the extent of 10,000 MW, by 2010. The technology is self-reliant in our country, and a major nuclear power plant is commissioned at Kudankulam.

Table 6.4 Percentage of nuclear energy generated

Country	Nuclear energy as % of total energy
France	76.10
Germany	29.10
Japan	33.40
South Korea	36.10
Sweden	46.60
Switzerland	39.90
United Kingdom	25.00
United States	22.50

In our country, atomic power stations have been set up in Tarapur (Bombay), Narora (Uttar Pradesh), Kota (Rajasthan) and Kalpakkam (Tamil Nadu).

Wind Energy

Seasonal winds carry energy that can be tapped, and in certain regions of the globe, during the major part of the year, a steady and regular gust of air exits. When the wind turbines are placed across the wind, energy could be generated. The wind-energy generators are common in USA, Germany and India. Low and high hub generators are in use. Traditionally, rural populations used windmills for the following applications.

◇ To perform agricultural tasks such as grinding corn, crushing sugar cane, threshing, and woodcutting.

◇ For special applications such as moving saline water in salt works.

◇ For the generation of electricity (in more recent times).

The use of wind as a source of the most widely utilized form of energy electricity began in the nineteenth century. Indeed, the windmills had a major role in sparking off the industrial revolution. Wind energy can be used to run a windmill which in turn drives a generator to produce electricity. Wind can also be used to provide mechanical power for water pumping. The potential of wind energy as a source of power is large. The energy available in the winds over the earth's surface is estimated to be 1.6×10^7 MW, which is of the same order of magnitude as the present energy consumption of the earth. However, the wind does not blow with required intensity all the year round and in all areas (Figure 6.2). Therefore, wind power can be used only in certain areas and on certain days.

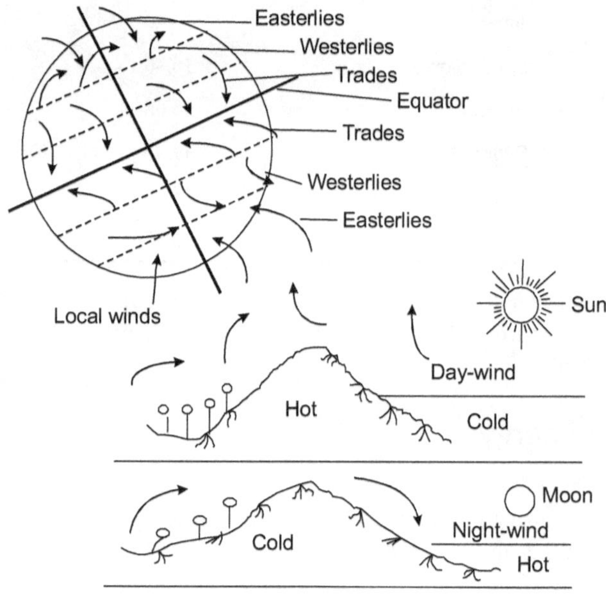

Figure 6.2   Day- and night-winds

Wind energy–Indian scenario   Wind energy generation has been given a strong thrust by Ministry of Non-Conventional Energy Sources, Government of India. Private sector entrepreneurs have also successfully marketed wind energy in our country.

Wind energy has proved to be an economically competitive renewable source of energy in certain locations in India. These locations include

- ◇   coastal regions in Gujarat, Kutch, and Saurashtra.

- ◇   gaps in the Eastern Ghats in Tamil Nadu, Tuticorin, and Kayathar.

- ◇   plains in Rajasthan, UP, MP and hilltops.

The total estimated wind energy resource potential in India is about 25,000 MW. Of this about 5,000 MW is located in Gujarat and 6,000 MW in Tamil Nadu.

In Tamil Nadu, the selected sites include Kayathar, Muppandal, Ayukudi, Alagiapandipuram, and Tuticorin. In Gujarat, Mandvi and Okha, wind farms are famous.

Energy from Seas

All flowing waters carry with them a kinetic energy. When such water encounters a turbine, part of the momentum of the flowing water is transferred on to the turbine, causing it to rotate. The rotation of the turbine can then be used to generate electricity. Whether the water is in the open ocean, an estuary, or a river, its motion can thus be utilized in generating energy. Thus, seas can be utilized as a source of wave, tidal or ocean thermal energy. UK and Japan are the pioneers in this area. About 13-KW per metre height of the wave can be generated. Ocean power utilizes the temperature gradient that exists down the ocean depths, i.e., between the warm, surface seawater at about 28°C and the cold, deep seawater at 5–7°C at a depth of 800–1000 m in tropical areas, and generates large quantities of power by the thermoelectric principle proposed by Thomson and Pettier. In some places, where the hot-water current is below the cold-water surface, an upwelling convective type of motion of the seawater can be used to run turbo pumps and this is on the drawing board with the scientific works on.

India's potential is large along the coastal length from Bombay to Vishakapatnam. The total potential from sea is estimated as 50,000 MW from ocean thermal energy conversion, 40,000 MW from wave energy, and 8,000 MW from tidal energy.

Tidal Energy

Tidal energy is the energy that can be tapped from sea. Tidal waves of the sea can be used to generate electricity. Tides are generated by the action of gravitational forces of the sun and the moon on the oceans, by the spinning of the earth around its axis and the relative positions of the earth, moon and the sun. The ebb and low tide phenomenon of the ocean where the gravitational effect of the moon causes the sea water level to be raised up and down. This effect is predominant in coves and sounds, and a surge of sea water level more than 30 metres are observed in certain specific coastal areas. The resulting huge water storage bodies during the course of day can be made to operate the power generators that work in both directions. Experimental plants are known to exist in USA as well as in the west coast of India. Despite the very high potential, the technology is still in the infancy.

The most attractive tidal power sites are the Gulf of Cambay and the Gulf of Kutch, where the maximum tidal range is of the order of 11 m and 8 m respectively, and the average tidal range is of the order of 6.8 and 5.2 m respectively. The techno-economic feasibility of the Gulf of Kutch Scheme was taken up by the Central Electricity Authority, Government of India. The scheme envisages a single basin, single effect development with an installed capacity in the range 800–1000 MW. The main tidal barrier is about 3.25 km. But this project was not taken up, as it was too expensive when compared to the conventional hydropower projects. Another potential site is the Sunderbans with a maximum tidal range of 5.0 m and an average tidal range of 3.0 m.

The estimated tidal power potential in India is about 15,000 MW. The promising sites for tidal power plants are located in Gujarat and West Bengal. A survey of other sites in Orissa, Tamil Nadu, Kerala, Karnataka and Maharashtra, Andaman Nicobar, etc. is being done by the Non-conventional and Renewable Energy Department. The sites have good prospects but require extremely high investments compared to conventional power plants of the same rating.

Geothermal Energy

Geothermal energy derives the heat in the centre of the earth, and it is stated that a huge potential exists in this form of energy (Figure 6.3). So far, India does not appear to have a major-exploitable source. Geothermal energy can also be used for cooling by using heat for vapour absorption. In some places, the heated water comes to the earth's surface as hot springs. It can be used for heating water and for generating electricity. The thermal springs provide hot water and steam, and this can be used to run the turbo generator to be utilized for captive generators. The usefulness of this source in our country is limited. Hot springs are found in the places near Himalayan valleys and some areas in Assam. The technology is well-known, and small quantities of power have been generated in the US.

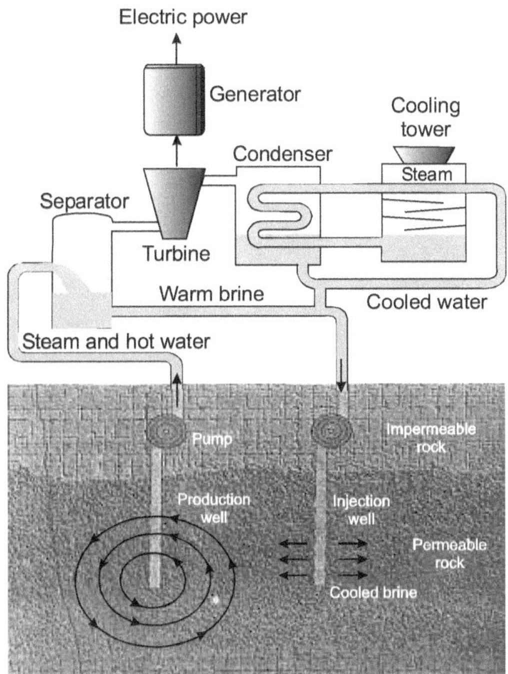

Figure 6.3 Geothermal energy production

Today the total installed geothermal power production in the world is only about 6,000 MW, a rather small quantity when one considers a single nuclear power plant that may exceed this level of power output.

Geothermal energy offers an eco-friendly source of electricity. Geothermal power plants meet the most stringent environmental regulations and release little, if any, carbon dioxide, a greenhouse gas suspected of contributing to global warming. Geothermal power plants are highly reliable and can operate twenty-four hours a day. Most power plants operate more than 95 per cent of the time. Hydrothermal reservoirs and earth energy are the two currently used forms of geothermal energy. To use the other three forms (namely, geo-pressured brines, hot dry rock, and magma), advanced technologies must be developed (Figure 6.4).

Iceland is well-known for its use of geothermal power. Two-thirds of its dwellings are heated geothermally. The steam from one of its large geothermal reservoirs has been used for heating the entire town of Hveragerdi, and a 17-MW power-generating plant was installed there in 1969.

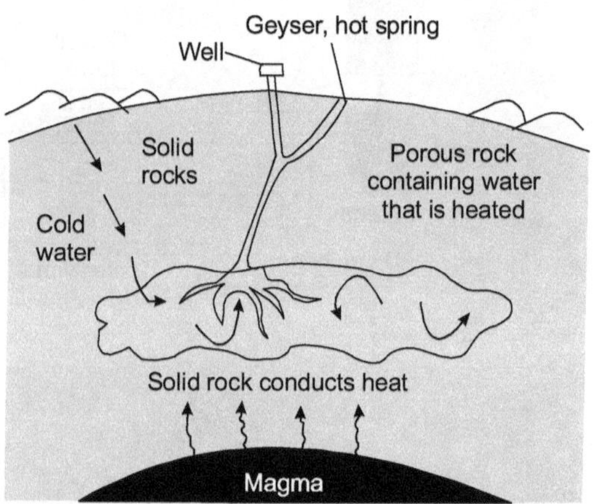

Figure 6.4   Model of a high-temperature hot-water geothermal system

In most of the houses of Reykjavik, icelands are heated by the hot water tapped from hot lava. In the Soviet Union, geothermal power is produced at three small plants in Kamchatka, with a total capacity of less than 40 MW.

In Japan, geothermal power production was begun at Masukawa in 1966 and at Otake in 1967. The capacity was 20 and 13 MW respectively in 1969, but later on, each was expanded to about 60 MW. Today, Japan has a total geothermal power production of about 215 MW.

Biomass and Biogas

Biomass is another renewable source of energy in the form of wood, agricultural residues, etc. The potential for agricultural residues alone is estimated at 480 MT with residues from food grains contributing about 100 MT. In India, the potential for application of biomass as an alternate source of energy is very great. We have plenty of agricultural and forest resources for production of biomass. Biomass is produced in nature through photosynthesis achieved by solar energy.

Biogas is another form of non-conventional sources of energy. Cattle dung and livestock dung are the main source for production of biogas. Cattle dung is widely used as a fuel in rural areas of our country. This deprives our fields of valuable organic manure. Now cattle dung is used in **Biogas** or **Gobar Gas Plant** to produce an odourless, low-pressure gas. This gas can be used for cooking and heating. The residue is used as manure.

The production of biogas is of particular significance for India because of its large cattle population. Some of the other sources of biogas are sewage wastes, crop residues, vegetable wastes, water hyacinth, poultry droppings, pig manures, alga ocean kelp. In the rural sector, biogas finds great applications in lighting, mechanical power and in the generation of small electricity.

Hydroelectric Energy

It is produced from the kinetic energy of water falling from a height. A number of power stations have been established on many rivers in our country.

As regards non-conventional sources of energy (renewable energy), its potential in our country is estimated at 100,000 MW. There has been a steady increase in power generation based on renewable resources.

THE NEED FOR ENERGY CONSERVATION

The situation of India's energy front is alarming, to say the least. There is already a 7% energy shortage of electric power. Just as there is about 55% gap between indigenous production of crude and demand for petroleum products.

India neither can afford the huge capital costs needed for installation of new owner plants, development of coal mines, strengthening of railway network, etc. nor can it afford the increasing drain on precious foreign exchange for oil imports. And, in any event, demand for energy is likely to continue to outstrip supply, at least for the next few years.

By preventing the wastage of energy and by efficiently using the available energy, we can add up to saving of 30%, which helps the country to face a major energy crisis.

The National Energy Conservation Movement has been launched to make consumers aware of their need to conserve energy. It provides simple rules and recommended energy-saving procedures. In homes, offices, factories and fields, one can save not only the energy but also the money with reduced electricity bills.

CONSERVATION OF ENERGY

The present critical energy position demands an organized effort at all levels from an individual to an international action. The conservation of energy is considered as a quick and an economical way to solve the problem of power shortage and as a means of conserving the country's finite sources of energy. The energy conservation measures are cost-effective, requiring relatively small investments and having a short gestation as well as payback periods. The studies conducted by Energy Management Centre, New Delhi have indicated that there is about 25% potential of energy conservation in the industrial sector. The following measures can help in this effort:

1. Development of technologies for the use of solar energy in appliances and transport vehicles.

2. Development of efficient and smokeless Chulhas or wood stoves.

3. Development of non-conventional energy sources and less dependence on fossil fuels.

4. Planned programme for raising fuel wood, trees and shrubs under the control and maintenance of local communities especially in developing countries.

5. Effective use of agricultural and animal wastes to obtain biogas and manure.

6. Improvement of engine and pump designs to increase the fuel efficiency.

7. Development of effective techniques to trap the wind and tidal energy.

The land is one of the most vital resources for mankind. It is used to produce agricultural crops and other biological materials needed for food, fodder, medicine, fibre, and so on. The availability and cultivability of land has influenced the pattern of energy use. The usable land is defined in terms of its topography and spatial nature. (Underground geological resources such as oil, gas, ores, precious metals, and deeper geo-hydrological resources normally bear no relation to the surface topography.) This definition is associated with an economic value of the resource.

Land is an area of the earth's terrestrial surface, which includes the near-surface climate, the soil and terrain forms, the surface hydrology (shallow lakes, rivers, marshes, and swamps, associated groundwater reserves), the plant and animal populations and the human settlement patterns and so on.

A natural unit of land hills covers both the vertical aspect, (from atmospheric climate down to groundwater resources) and the horizontal aspect, (an identifiable repetitive sequence of soil, terrain, hydrological, and vegetative or land use elements). Thus, the components of the natural land unit are termed the land resources, including physical, biotic, environmental, infrastructural, social and economic components, in as much as they are fixed to the land unit.

LAND RESOURCES IN INDIA

Of India's total geographical area of 328.73 million hectare (mha), only 264.5 mha is under use for agriculture, forestry, pasture and other biomass production. Since 1970/71, the net area sown has remained around 140 mha and was 142.22 mha during 1998/99 (Ministry of Agriculture and Cooperation, 1992).

India has a total land area of 2.4 per cent of the world's total and supports a population over 15 per cent of the world. The per capita land availability is 0.48 hectare, as against 8.43 hectare in the former Union of Soviet Socialist Republics and 0.98 hectare in China. With an increase in industrialization, urbanization and infrastructure, development is progressively taking away considerable areas of land from agriculture, forestry, grasslands and pastures.

LAND DEGRADATION

The disturbance to physical, chemical, and biological features of the soil that alters the socio-economic features of the area is termed as land degradation. The surface layer of land is called the soil. It covers about four-fifth of the land areas. The fertility or the productive capacity of the soil depends on the minerals it contains. As the minerals abound in the top layer of the soil, the top layer is the best for vegetation. Some factors deplete the mineral content of the top layer and damage it, thereby reducing the fertility and productive capacity of the soil. They are:

    i.   Soil erosion

   ii.   Soil pollution

  iii.   Salination and waterlogging

  iv.   Shifting cultivation

   v.   Desertification

  vi.   Urbanization

Soil Erosion

Soil erosion is the wearing away, detachment and transportation of soil from one place to another place, i.e., movement of soil components, especially surface litter and topsoil, from one place to another. It results in the build-up of sediments and sedimentary rocks on land and in bodies of water. The two main agents of erosion are flowing water and wind, with moving water causing most soil erosion. (The association between water and land is so intimate that the water element cannot be excluded because land as a unit "intermixed" with water. Also, it affects the quality and quantity of the passing water.) Soil and water are the most essential requirements for the growth and sustenance of all lives.

Soil erosion results from natural and human activities. In undisturbed vegetated ecosystems, the roots of plants help anchor the soil, and usually soil is not lost faster than it forms. Soil becomes more vulnerable to erosion through activities that destroy plant cover, including farming, logging, construction, overgrazing by livestock, off-road vehicle use, and deliberate burning of vegetation.

Such human activities can speed up the erosion and destroy in a few decades what nature took hundreds to thousands of years to produce. The National Bureau of Soil Survey and Land Use Planning data show that nearly 3.7 million ha suffer from nutrient loss and/or depletion of organic matter. The problem is widespread in the cultivated areas of the subtropical belt, including areas under shifting cultivation or jhuming in the northeastern states. According to the Government of India, the quantity of nutrients lost due to erosion each year ranges from 5.8 to 8.4 million tons.

When surface soil, denuded of their natural protective cover by overgrazing of grasses, excessive ploughing, and destruction of forests is exposed to wind and water, heavy erosion is caused by removal of the top soil.

Damages caused by erosion   The two major harmful effects of soil erosion are loss of soil fertility and loss of its ability to hold water and sediment. Sediment run-off can pollute water, kill fish and shellfish, and clog irrigation ditches, boat channels, reservoirs, and lakes.

- ◇ Removal of the topsoil decreases the water-holding capacity of the soil and hence productivity goes down.

- ◇ Low-lying areas are exposed to the danger of deposition of coarser particles, which are washed from higher hilly areas, making the soil less productive.

- ◇ Tanks are filled every year during monsoon season by water from catchment area. This water also brings with it large quantities of silt and clay.

If proper care is not taken, reservoirs are silted and their storage capacity is considerably reduced.

- ◇ If surface run-off is allowed to go on unchecked, the quantity of water that should infiltrate into the soil is very much decreased.

Off-site impacts of erosion   Higher erosion rates have resulted in the sedimentation of river-beds, and siltation of drainage channels, irrigation canals, and reservoirs. Siltation has changed the hydrology of several watersheds of the country, resulting in a greater frequency and severity of floods, and reducing the water availability in dry season. The siltation rate of reservoirs in India has been estimated to be much high than the values assumed at the time of designing. This has drastically reduced the life of projects, which involved huge investments.

Flooding   Increased sedimentation and reduced capacity of drainage systems. Increased gully erosion and ravine formation results in increased run-off and peak discharge for any given rainfall from watersheds.

Satellite imagery of Himalayan torrent shows that between 1990 and 1997 the width of torrents has increased by 106% and that of rivers by 36%. Consequently, streams and rivers overflow their banks, flooding the downstream areas.

The problem of human-induced waterlogging in India is more common in canal command areas (surface irrigation) because irrigation facilities are often introduced without adequate provision for drainage resulting in a rise of the water table.

Global effects of soil erosion   Following points enumerate the seriousness of soil erosion.

- ◇ A 1992 joint survey by the United Nations (UN) Environment Programme and the World Resources Institute estimated that topsoil is eroding faster than it forms on about 38% of the world's cropland.

- ◇ The survey also found that 15% of the world's land (two- thirds of it in Asia and Africa) was degraded to some extent by soil erosion.

- ◇ Recent studies show that in northwest region of China, a combination of overploughing and overgrazing is causing massive wind erosion of topsoil. At times, the eroded soil rises in huge dust plumes that blot out the sun and reduce visibility in China's northeastern cities.

- ◇ The plumes also reduce visibility and increase air pollution in Japan, the Korean Peninsula, and the northwestern United States.

◇   Nearly 40% of the world's land (75% in Central America) used for agriculture is seriously degraded by erosion, salt build-up (salinization), and waterlogging.

◇   Soil degradation has reduced food production on about 16% of the world's cropland. The situation is worsening in some developing countries as many poor farmers depend up on marginal (easily erodable) lands to survive.

◇   According to a soil expert's estimation, soil erosion causes damages of at least $375 billion per year (an average of $42 million per hour) worldwide, including direct damage to agricultural lands and indirect damage to waterways, infrastructure, and human health.

◇   Soil erosion is not a high priority for many governments and farmers because it usually occurs so slowly that its cumulative effects may take decades to become apparent.

◇   For example, the loss of 1 millimetre (0.04 inch) of soil is so small that it goes undetected. However, over a 25-year period, the loss would be 25 millimetres, which would take about 500 years to replace by natural processes.

◇   Soil erosion is very extensive in Australia, India, Central America, China, Nepal, USA, Spain, and Russia.

◇   In India, mainly due to the overgrazing of the livestock, there is maximum soil erosion of topsoil contributing to 18.5% of the global loss. It is estimated that strong flooding in Shivalik hills removes 6 cm of fertile topsoil, in one year, which took 2,400 years for its ecological construction.

◇   Extreme soil erosion by wind and water has formed huge sandy tracks in Punjab, Haryana, and Chambal valley. The 7000-year-old Thar Desert of Rajasthan was once a seacoast rich in vegetation and civilization.

◇   Erosion, floods and sedimentation result in siltation and consequent clogging of irrigation canals due to deposition of silt in the streams and rivers.

Although soil erosion is a serious problem, some analysts argue that some erosion estimates are overstated. They contend that some surveys underestimate the abilities of some local farmers to restore degraded land. Some analysts also contend that much of the eroded topsoil does not go far and is deposited further down a slope, valley, or plain. In some places, the loss in crop yields in area could be offset by increased yields elsewhere.

SOIL AND WATER CONSERVATION METHODS

Soil is important as it provides foothold for the plants and majority nutrients needs by them. Along with soil, water is another important factor essential for all life and production of food. If rainwater is not conserved properly, it will not only cause scarcity and famine but also washes away the soil, which is a valuable national asset. It is therefore the prime responsibility to conserve soil, which is the main capital of the farmer as well as the nation, at all costs.

Recognizing the seriousness of erosion problem, the central government established the Central Board of Soil Conservation to assist the states and River Valley Projects. It takes centuries to form one-inch layer of soil but does not take long to lose it by erosion. Soil, especially topsoil, is classified as a renewable resource because natural processes regenerate it. However, in tropical and temperate areas it takes 200–1,000 years (depending on climate and soil type) for 2.5 cm (1 inch) of new topsoil to form. If topsoil erodes faster than it forms on a piece of land, it eventually becomes a non-renewable resource.

To minimize the loss of soil and water and to cultivate land without much harm to the soil, the following agronomic and mechanical measures are followed.

Agronomic Measures

Strip cropping This consists of growing erosion-permitting crops, e.g. cotton, jowar, bajra, etc., in a strip and erosion-resisting crops, e.g. groundnut, matki, hulga (*Dolichos biflorus*), soybean, in an alternate strip. The soil flowing from the strips growing erosion-permitting crops is caught by the alternating strips of erosion-resisting crops.

Mulching It is a natural or artificially applied layer of plant residues or other materials on the surface of the soil with the objectives of moisture conservation, reduction of run-off and erosion and soil losses, e.g. jowar or bajra stubbles, paddy straw or husk, sawdust, etc.

Crop rotation   Growing a set of crops in a regular succession over the same field within a specified period is termed as rotation. When continuously grown, jowar or bajra crop causes more erosion, but if followed by a legume crop namely hulga, matki or gram that covers the soil, it causes less erosion.

Contour cultivation   Tillage operations such as ploughing, sowing and intercultural should be done across the slope of land. This will help creating obstruction to the flow of water at every furrow, which acts like a small bund and will result in uniform distribution of water, and less run-off and erosion.

Planting of grasses for stabilizing bunds   Grasses prevent the soil erosion and improve the soil structure. Several grasses as well as legumes were tried on bunds should give the maximum root growth and canopy coverage and stabilize bunds effectively, e.g. anjan, marvel-8, rhodes, thin napier, blue panic, kusal.

Planting of trees and afforestation   Forests conserve soil and water quite effectively. They not only obstruct the flow of water, but also provide organic matter, which increases the water-holding capacity of soil.

Mechanical Measures

These measures require engineering techniques and structures. Bunding-Tals, i.e., big bunds across large blocks of sloping lands, constructed of earth or stone or both, can impound water and arrest soil washed from the fields lying above.

Contour bunding   It consists of construction of a series of earthen bunds of suitable sizes along contours at a lateral distance of every 60 m, or a fall of 1 to 1.5 m. The slope of land is thus broken into smaller and more level compartments, which hold soil as well as rainwater. The size, cross-section and interbund spacing depend upon the nature of rainfall, soil and slope of the area.

Graded bunding   In high-rainfall areas, drainage of surplus water has to be attended to avoid waterlogging of soil. The bunds are therefore slightly graded longitudinally for the safe disposal of water. For the safe removal of excess run-off water, it is essential to provide suitable outlet structures at proper places so that no damage is done to bunds, e.g. stone outlets, channel weirs or pipe outlets in low rainfall area.

Terracing   It is suitable on a bigger slope up to 10% and rainfall is higher than 1250 mm. Terrace bunds consist of comparatively narrow embankments constructed at intervals across the slope, and the vertical spacing between bunds may vary from 1 to 2 m, depending upon the slope, types of soil, rainfall, etc. Bench terracing is done when gradient is steeper than 10% as in hilly ranges of Himalayas, Sahyadri, etc.

Control of stream and river banks   River banks should be protected by providing spurs, jetties, rivets and retaining walls. Adjoining areas should be stabilized under permanent vegetation.

Shifting cultivation   Shifting cultivation with a short fallow cycle does not allow enough time for the land to recuperate naturally and is responsible for large-scale (about 4.5 mha) land degradation in several parts of the country. The practice, a socio-economic outlet, needs to be discouraged and alternatives to the people engaged in the practice need to be provided in a phased manner for their livelihood.

Desertification

Increasing human population has put a great pressure on the land. Vast areas of land have been cleared for the cultivation of agricultural crops.

The destruction of natural vegetation results in the accelerated soil erosion due to the removal of the vegetation cover. Erosion of the top fertile soil results in the loss of soil productivity and in the formation of deserts. Wind causes the shifting of sand dunes from one place to another, increasing the proportion of fertile land being converted into desert each year.

Desertification is a complex process that involves multiple natural and human-related causes and that continues at varying rates in different climates. It results mainly from a combination of prolonged drought and unsustainable human activities.

It is formed by destroying thousands of hectares of productive land. Excessive grazing by livestock is another factor resulting in desertification, especially in Rajasthan. All efforts to check the advancement of deserts by afforestation are of no use because large herds of goats and other animals destroy the saplings planted. Many deserts in the world are a result of human activities.

The process of desertification   In desertification, the productive potential of arid or semiarid lands falls by 10% or more because of a combination of natural climatic changes, which causes prolonged drought and human activities that reduce or degrade topsoil. The process can be moderate (with a 10–25% drop in productivity), severe (with a 25–50% drop), or very severe (with a drop of 50% or more, usually creating huge gullies and sand dunes). Desertification is a serious problem in many parts of the world (Figure 7.1).

Figure 7.1   Desertification

Desertification at global level   An estimated 8.1 million square kilometers—an area the size of Brazil and 12 times the size of Texas—have become desertified in the past 50 years. About 40% of the world's land and 70% of all dry lands are suffering from the effects of desertification. In addition, each year about 150,000 square kilometers (58,000 square miles)—an area larger than Greece—becomes desertified. This threatens the livelihood of at least 135 million people in 100 countries.

Managing and combating desertification and drought   Desertification is land degradation in arid, semi-arid and dry sub-humid areas, caused by many factors, including climatic variations and human activities. It affects about 1 billion people and one quarter of the world's total land area. Desertification is most severe in the world's dry lands amounting to about 3.6 billion hectares. The decline in productivity, and loss of crop and livestock production in these areas, has resulted in widespread poverty, hunger and malnutrition.

In combating the desertification and drought, national governments and the international community should aim to strengthen the knowledge base and to develop information and monitoring systems for all areas prone to desertification drought that incorporate the economic and social aspects of these vulnerable habitats. Adequate worldwide systematic observation systems are helpful for the development and implementation of effective anti-desertification programmes. An integrated and coordinated information and systematic observation system based on appropriate technology is essential for understanding the dynamics of desertification and drought processes in all levels, i.e., national, global, regional, and local levels.

- ◇ Preventive measures should be launched in non-affected and slightly affected areas; corrective measures should be implemented to sustain the productivity of moderately desertified land; and rehabilitative measures should be taken to recover severely desertified dry lands.

- ◇ Current livelihood and resource use systems cannot provide adequate living standards in these fragile areas experiencing drought and increasing population pressure. Poverty has, to a major extent, accelerated the rate of environmental degradation and desertification.

- ◇ Action is needed to rehabilitate and improve subsistence agriculture and agro-pastoral systems for sustainable management of rangelands as well as the promotion of alternative livelihood systems.

- ◇ Developing comprehensive anti-desertification programmes and integrating them into national development and national environment planning are essential. In a number of developing countries, the natural resource base subject to desertification is the main resource upon which the development depends.

- ◇ Efforts should be made to develop comprehensive drought preparedness and relief schemes, including self-help arrangements, for drought-prone areas, and to design programmes to cope with environmental refuges.

◇ Early warning systems, which forecast drought, will make possible the implementation of drought preparedness schemes. Integrated programmes at local levels, such as alternative cropping strategies, soil and water conservation, and the promotion of water harvesting techniques, could reduce the impacts of drought while providing basic necessities.

◇ Encouraging and promoting popular participation and environmental education, focusing on desertification control and management of the effects of drought, is also crucial.

◇ Whether the projects and programmes related to desertification and drought control, to date, succeed or fail, the experience points to the need to focus on obtaining popular involvement rooted in the concept of partnership and on the sharing of responsibilities by all parties.

◇ Watershed management programmes have been taken up extensively in the recent past.

◇ The Soil and Water Conservation Division in the Ministry of Agriculture has been playing a key role in implementing integrated watershed management programmes with a plan to cover 86 mha.

◇ Over 30,000 hectares of shifting and semi-stable sand dunes have been treated with shelterbelts and strip cropping.

◇ All India Soil and Land Use Survey is engaged in generating spatial and non-spatial information on the soils of India and in preparing thematic maps like land capability classification, hydrological soil grouping, irrigability classification, etc. The state governments are also working on various aspects of soil conservation following the guidelines of the centre.

The degradation of soil conditions is probably the most important issue since it affects a major life-supporting system and because its natural recuperation may take centuries. Artificial soil rehabilitation is often very expensive. It should be noted that increased rural population in marginal areas does not necessarily lead to land degradation so long as improved technologies and policies are implemented. According to the expert soil managers, if basic production facilities, primary education, availability of rural credit for integrated soil–water–nutrient conservation, plus access to a nearby market for agricultural products are improved, then higher population densities may naturally lead to intensified land use in a sustainable form.

LANDSLIDES

Landslides are formed when rock, earth, or debris flows on slopes due to gravity. They can occur on any terrain given the right conditions of soil, moisture, and the angle of slope (Figure 7.2). Being integrated to the natural process of the earth's surface geology, landslides serve to redistribute the soil and sediments in a process that can occur in abrupt collapses or in slow gradual slides. Such is the nature of the earth's surface dynamics. Also known as mud flow, debris flow, earth failure, slope failure, etc., a landslide can be triggered by rains, floods, earthquakes, and other natural causes as well as human-made causes, such as land grading (to achieve proper drainage), terrain cutting and filling, excessive development, etc. (Figure 7.3).

Figure 7.2   Typical landslide

Factors affecting landslides can be geophysical or human-made; they can occur in developed areas, undeveloped areas, or any area where the terrain was altered for roads, houses, utilities, buildings, and even for lawns in one's backyard.

Figure 7.3   Induced landslide

The US Geological Survey, working with other federal agencies, has efforts underway to study, plan, and mitigate landslide risks, so have some communities across the country. Many deal with landslides as part of flood control, erosion control, hillside management, earthquake hazard mitigation, road stabilization, and other programmes.

Effects of Landslides

◇   Perhaps the most common reminders of landslide risks are those "Watch for Falling Rocks" highway signs. Although "sliding rocks" is more apt, very few can see a landslide. Occasionally we see small rocks or debris on the pavement, but a large size slide usually starts with such small incidents.

◇   Visually, a landslide resembles a snow avalanche, only with a louder rumbling noise, and is capable of generating enough force and momentum to wipe anything in its path. One such devastating landslide wiped entire towns and villages in Columbia in 1985 and killed about 20,000 people.

◇   Figure 7.4 shows the recent landslide. They show what has left after a slide. In some cases, only the rail or pavement is mangled, whereas in others a house or building crushed. However, in almost every aftermath, the losses are real, the damages are total, and the terrain changes are permanent.

Figure 7.4   Road block due to the landslide

◇   Landslides cause one to two billion dollars in damage each year in the US and claim as many as fifty lives. Although damage estimates vary by how one classifies disasters, it is significant.

◇   Landslides affect utilities, transportation, and all other forms of infrastructure, whether public or private.

◇   Pittsburgh and Cincinnati are two examples of urbanized areas with frequent landslides, where developments on hills and hillsides are common. In the Great Plains, heavy rains combined with loss of vegetation due to wildfires trigger landslides in clay-rich rocky areas.

◇   On the west coast, earthquakes add to the causes of landslides. For example, the 1994 Northridge earthquake triggered many thousand landslides in the Santa Susanna Mountains. In short, no region of the country is safe from landslides, whether caused by geophysical or human-made factors.

◇   Most landslides delivered large quantities of sediment to stream channels and many travelled as "torrents" or mud flows for hundreds or thousands of feet down the hill slopes and stream channels before coming to rest in larger streambeds and river channels (Figure 7.5). It is estimated that over 80% of the observed landslides delivered the sediments directly to stream channels.

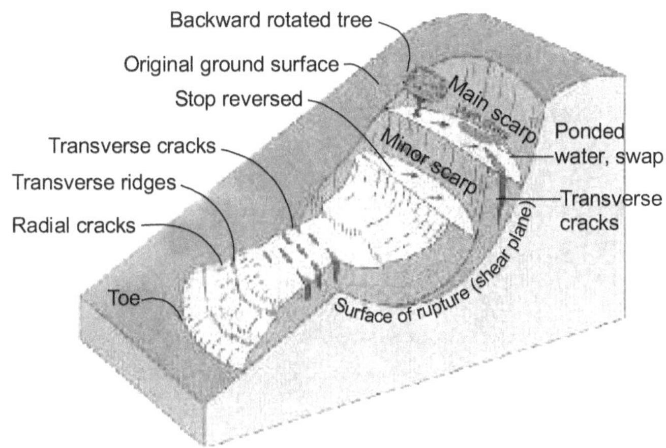

Figure 7.5   Movement of landslide

Remediation to Avoid Land Slides

◇ Removal of vegetation from steep swales in watersheds, where debris tormenting is a common land sliding process, will greatly increase the number of such landslides that occur during a large storm.

◇ Construction of roads by side casting on steep slopes is hazardous and will result in greatly increased rates of land sliding and stream sedimentation.

◇ When the next storm occurs (whether or not it is of lesser or of greater magnitude), additional land sliding and erosion will occur in these watersheds.

◇ The impacts of coarse sediments introduced into streams and rivers by land sliding and road failures may persist for decades, depending on transport rates.

◇ The impacts to lower, larger streams and rivers may actually increase over the near term (next several years), as sediments from the headwater areas is moved downstream and is deposited in lower gradients.

◇ Restoration measures for the sediments already delivered to the stream system by landslides and road failures will be impractical and not cost-effective.

Prevention of landslides from harvested (clear cut) slopes is dependent on the recognition and avoidance of sites with high-risk characteristics. For example, clear-cutting some watershed areas exhibiting steep headwater swales appears to have dramatically increased the landslide activity. Once cut, the slopes are vulnerable for a period of years until they are well vegetated. Clear-cut harvesting is best avoided in sensitive watersheds or where downstream aquatic resources are threatened.

CONTROL OF LAND DEGRADATION

◇   A well-defined integrated land use policy at the implementable level should be developed at the earliest, as also rural fuel wood and grazing and fodder policies to scientifically and sustainably manage the land and the forests.

◇   Indian conditions, which along with scientifically sound land management practices, would address land degradation problems and maintain land quality for sustainable use.

◇   As far as possible, land should be managed on a natural watershed basis. Because it presents an ideal unit for the most effective management like increasing the utilization of irrigation potential, promoting water conservation and efficient water management along with expansion of irrigation facilities.

◇   A correct assessment of the nature and extent of the existing degraded land through rapid inventory remote sensing techniques and GIS, needs to be carried out as early as possible with scientifically sound criteria and indicators.

◇   Soil nutrient mining results in serious soil health and ecological problems, which needs urgent attention. Integrated plant nutrient system (IPNS) has to be adopted to improve the fertilizer use efficiency and to reduce the potential danger of pollution from higher nutrient use in agriculture.

◇   A systematic monitoring mechanism needs to be developed to assess the balance between input and withdrawal of nutrients to guard against possible nutrient depletion.

◇   Domestic and municipal wastes, sludge wastes, pesticides, industrial wastes, etc. must be used with utmost caution to avoid the possibility of pollution of soil through heavy metals and other toxic substances which are often present in them.

◇ Shifting cultivation with a short fallow cycle does not allow enough time for the land to recuperate naturally and is responsible for the large-scale (about 4.5 mha) land degradation in several parts of the country.

◇ Land shortages and poverty farther lead to non-sustainable land management practices, and is one of the most important causes, and effect nexuses of land degradation. In this context, the participatory approach needs to be adopted to mitigate some of these adverse effects.

◇ To harmonize developmental activities such as industrialization, urbanization, mining and infrastructure development, make them compatible with surrounding land use and guard against any form of land degradation.

Education, training, research, and technology development would enable us to focus on analysing and adapting conditions and principles for sustainable land use as well as resource conservation technologies and practices.

INTRODUCTION

Conservation of resources is the management of the human use of natural resources to provide the maximum benefit to current generations while maintaining capacity to meet the needs of future generations. Conservation includes both the protection and rational use of natural resources.

Earth's natural resources are either non-renewable (minerals, oil, gas and coal) or renewable (water, timber, fisheries and agricultural crops). Natural resources are the basic goods and services that sustain human societies. Renewable resources play central roles in providing air, water, and food. Non-renewable resources provide the energy essential for industrial economics and are the sources of important products ranging from iron tools to silicon chips.

Although humans are entirely dependent upon earth's rational resources, the combination of growing populations and increasing levels of resource consumption is now degrading and depleting the natural resources. The global population has grown to nearly seven times as large (6 billion) and the level of consumption of resources is far greater. This human pressure now exceeds the carrying capacity of many natural resources.

The best practices in the conservation of resources include the following:

1. There should be a minimum entitlement to human needs by virtue of being born into the human family, no matter in which country of the world.

2. Affluence should be regulated so that no one individual or group can take more than a fair share from the finite pool of world resources.

3. There should be a limit to damage that people can cause to the physical environment. For example, air and water pollution, stripping of the land, thinning of the ozone layer, inducing the greenhouse effect, etc.

4. Family-planning decisions, traffic accidents, weather modification and genetic engineering, need to be controlled to eliminate or at least to minimize the risks.

5. The rate of using non-renewable resources such as fossil fuels and hard minerals should be curtailed.

6. One should avoid the practices that affect the renewability of renewable resources (soil erosion, destruction of wildlife, over cropping of farmland, denudation of forests, over fishing of lakes and oceans) so that future generations will not live in deprivation.

7. Cultural identity should be preserved by improving the ideologies or social systems of any one group on others.

With an ever-increasing population, there is a rise in resource use. The ability of biological systems to support human populations is being seriously threatened. Hence, we should endeavour to judiciously use our natural resources and then pave the way for sustainable development.

Whether renewable or non-renewable, human beings have access to energy resources and manipulate the environment to their advantage.

SUSTAINABLE DEVELOPMENT

Earth has limitations and the natural resources are only limited. What is taken from the nature should be returned for its survival?

Sustainable Development is the process of social and economic betterment that satisfies the needs and values of all interest groups, while maintaining the future requirements of the available natural resources and diversity. It incorporates all the principles of ecological, social, cultural and economic sustainability.

Ways of Sustainable Development

◇ Stop imitating the high-energy-consuming and resource-depleting western pattern of development.

◇ Stop developmental patterns based on loans, subsidies imported resources and borrowed techniques.

◇ Move towards an eco-development using indigenous materials suited to our culture and traditions at low costs.

  ◆ Reduce the annual emission of CO_2, SO_2, and NO_2.

  ◆ Shift from fossil fuel to solar, wind, mini-micro-hydel and biogas.

  ◆ Reduce the energy consumption and improve the efficiency by efficient use of technologies.

  ◆ Support the economic development with conservation of ecological processes and biological diversity.

  ◆ Involve in community participation.

  ◆ Retain as much natural environment as possible.

We should help for the basic rights of women and children and the basic rights of the poor, weaker and indigenous sections of the society.

  ◇ There should be involvement of people in representation and participation of environmental issues.

  ◇ People should check population growth.

  ◇ There should be only the least damage to environment and species.

  ◇ The culture and values should be maintained by all.

  ◇ Public awareness should be heightened regarding the impacts of violence in society.

As an individual, one can really do anything significant to help preserve the earth.

  ◇ By developing, recycling, reusing and refilling attitudes, e.g. sort out garbage to separate items such as glass and paper that can be recycled. Bottles can be refilled. Learn to reuse out-of-date mails, bills, notices, etc. into notepads.

- Using a travel cup in your car instead of disposable styrofoam cup.

- Saving energy whenever possible, e.g. switch off the lights and fans when not necessary.

- By replacing incandescent light with energy-efficient fluorescent light and avoiding colour bulbs.

- Check whether the thermostat and regulator of your electric appliances are functioning properly to prevent unnecessary loss of energy.

- Run your dishwasher, washing machine and dryer only when you have full loads.

- Spending least amount of water in our lawn.

- Keeping the noise levels from radios, television sets or other household machines to a minimum.

- Buying stationery, cards and wrapping papers from recycled paper.

- Avoid buying "over-packed" goods and foods.

- Checking on pollution and informing local authorities wherever it occurs.

- By keeping the bus stand and railway station clean by providing them with bins and comfort stations.

- One may buy recyclable bond and computer paper.

- Students should be informed and regularly updated on local, regional, national and global environmental issues.

There is still a considerable lack of awareness of the interrelated nature of all human activities and the environment, due to inaccurate or insufficient information.

Developing countries, in particular, lack relevant technologies and expertise. There is a need to increase the public sensitivity to environmental and developmental problems and the involvement in their solutions, and foster a sense of personal environmental responsibility and greater motivation and commitment towards sustainable development.

EQUITABLE USE OF RESOURCES

Environmental changes normally occur over geological time under the influence of natural forces. However, increasing human activities have put an additional load on the top of these natural forces, considerably stressing the environment and threatening the earth's fertility.

Land Resources

Growing demands on land resources have led to soil erosion, salination and waterlogging, desertification, deforestation, and the disruption of many precious and economically vital ecosystems. The pollution and over use of the world's finite freshwater resources threaten all the socio-economic sectors. An integrated approach to land use should be used at all stages of decision-making, from goal-setting implementation. Legislation, regulations and economic incentives should encourage the rational use of land. Research is also important to determine the capacity of land and the interaction among various land uses and environmental processes.

Energy Efficiency

Depending largely on renewable fossil fuels as our primary source of energy has contributed to air pollution, acid rain, greenhouse warming, marine pollution and other adverse impacts. These pollutants in the environment poses a threat to human health and damages the forests, lakes and rivers and the fragile ecosystem. However, it is unlikely that fossil fuels will be abandoned in the short- and medium-term transitions. Improvements can be made in energy efficiency while making a long-term transition to environmentally safe and sound energy systems.

Food Resources

Freshwater is a finite resource and, in many parts of the world, is becoming increasingly scarce. Also, deforestation, urbanization, and poor farming and mining practices cause sedimentation in reservoirs. Excessive use of agricultural chemicals, acid rain from industrial pollution, dumping of untreated sewage and factory wastes, and over-pumping of groundwater all contribute to the deteriorating quality of water resources. An estimated 80 per cent of all diseases in the developing countries and one-third of the deaths are related to contaminated water. World food production must be more than double over the next four decades to meet the needs of a growing population. Since more than 80 per cent of the global population lives in the developing world, most of the new production will need to occur there. The challenge is to apply the most efficient possible food production methods to potential lands while drawing destructive agriculture practices away from marginal areas.

Forest Resources

Forest resources are essential to both the development and the preservation of the global environment. Mismanagement of forests is linked to the degradation of soil and water, loss of wildlife and biological diversity, and pollution and global warming. Each year some 17 mha of tropical forests are lost, as a result of agricultural and industrial expansion, overgrazing, excessive or poorly managed tree-cutting for lumber or fuel purpose and similar human pressures. Meanwhile, air pollution and fires are depleting the wooded lands of many developed countries. A holistic approach to forest conservation and development must address all the relevant issues, including population pressures, unsustainable agricultural and industrial practices, land ownership, employment opportunities and external debts.

Desertification affects one-fourth of the earth's land areas and one-sixth of its people. In order to combat desertification, improved land and water use and reforestation are critical. An increased research and the provision of alternative livelihoods for subsistence farmers and herders are also needed, with more emphasis on preparing for drought emergencies. The rapid deterioration of mountain ecosystems threatens the planet's biological diversity and the well-being of many people. Proposals focus on halting the erosion and replanting the damaged areas, making natural disaster plans, and offering alternative employments to people whose livelihoods are linked to harmful environmental practices.

Coastal Settlement

Coastal and island settlements are threatened by human activities such as fishing, shipping, tourism, urban waste, and pollution caused by industry, agriculture and forestry. Half of the world's population lives less than six kilometres from the sea. By the year 2020, three-fourths may reside in this coastal zone, which today includes many poor, densely crowded settlements. People's well-being is closely related to the condition of fragile coastal environments. The environmentally distinctive practices in coastal regions have to be curbed and damaged areas have to be restored.

SUSTAINING BIODIVERSITY

Sustaining the diversity of biological species—plant, animal and insect life—is a key element in the sustainable development of planet. Today, however, short-term economic development rarely considers conserving biodiversity, and usually works against it. At present, national and international policies, as well as market and accounting practices, do not encourage sharing the benefits of biodiversity. Instead they equate economic gain with resource depletion. Commitments to conserve biodiversity should be included in national development policies. This would involve policies influencing technology transfer, debt, trade and environmental accounting.

Conserving the world's natural resources and its biodiversity and protecting the overall biosphere, while at the same time substantially increasing their productive yield, is one of the greatest challenges of a civilization. Halting and then reversing this environmental degradation necessitates the mobilization of both progressive and valuable indigenous technologies. It requires relevant information and alternative approaches in the energy, agriculture, forestry and water sectors. It calls for increased awareness and training for more responsible resource use, and efforts to ensure that resource conservation does not threaten those who are dependent on these resources for their livelihoods.

The critical need is to adopt an anticipatory and precautionary policy approach to sustainable resource utilization. Only through such integrated, long-term, participatory policy planning and management strategies, a better quality of life for both present and future generation can be assured.

Elimination of Desertification

Desertification is land degradation in arid, semi-arid and dry sub-humid areas, caused by many factors, including climatic variations and human activities. The decline in productivity and loss of crop and livestock production in 3.6 billion hectares of desertified land has resulted in widespread poverty, hunger and malnutrition of 1 billion people.

In combating desertification and drought, national governments and the international community should aim to strengthen the knowledge base and to develop information and monitoring systems for all areas prone to the desertification and drought, which incorporate the economic and social aspects of these vulnerable habitats. Global assessments of the status and rate of desertification should take into account the socio-economic causes and their interactions with climatic cycles and drought. Adequate worldwide systematic observation systems are helpful for the development and implementation of effective anti-desertification programmes. An integrated and coordinated information and systematic observation system based on appropriate technology and embracing global, regional, national and local levels is essential for understanding the dynamics of desertification and drought processes.

In combating land degradation through inter alia intensified soil and water conservation, afforestation and reforestation activities, preventive measures should be launched in areas which are not yet affected or are only slightly affected by desertification; corrective measures should be implemented to sustain the productivity of moderately desertified land; and rehabilitative measures should be taken to recover severely or very severely desertified dry lands.

There is an urgent need to develop and strengthen integrated development programme, to eradicate poverty and promote alternative livelihood opportunities in areas prone to desertification and drought. Current livelihood and resource use systems cannot provide adequate living standards in these fragile areas experiencing drought and increasing population pressure. Poverty has, to a major extent, accelerated the rate of environmental degradation and desertification. Action is needed to rehabilitate and improve subsistence agriculture and agro-pastoral systems for sustainable management of rangelands as well as the promotion of alternative livelihood systems.

Developing comprehensive anti-desertification programmes and integrating them with international development and national environmental planning is essential. In many of developing countries, the natural resource base subject to desertification is the main resource upon which development depends. The social systems interacting with land resources make the problem much more complex and necessitate an integrated land planning and management approach. Strengthening international cooperation through the preparation of an international convention to combat desertification in all affected areas of the world, particularly in Africa, will be considered for further negotiation at the Rio Conference.

Efforts should be made to develop comprehensive drought preparedness and relief schemes, including self-help arrangement for drought-prone areas and design programmes to cope with environmental refugees. Droughts in differing degrees of frequency and severity are a recurring problem throughout much of the developing world, particularly in Africa.

Apart from the toll on human lives, the economic costs of drought-related disasters are high, due to lost production and diversion of scarce development resources drought will make possible the implementation of drought preparedness schemes. Integrated programmes at local levels, such as alternative cropping strategies, soil and water conservation, and the promotion of water-harvesting techniques, could reduce the impacts of drought while providing necessities.

Encouraging and promoting popular participation and environmental education, focusing on desertification control and management of the effects of drought, is also crucial. The experience to date as to whether projects and programmes related to desertification and drought control succeed or fail points to the need to focus on obtaining popular involvement rooted in the concept of partnership and the sharing of responsibilities by all parties.

EDUCATION, PUBLIC AWARENESS AND TRAINING

Education is the social institution entrusted with the main responsibility for passing on the wisdom, knowledge and experience to the succeeding generations. It represents a guided path, which helps individuals to understand their own societies and to take their place in them. Education is perhaps the single most important influence in changing the human attitudes and behaviour, in promoting the economic growth and raising the quality of life, and in providing the knowledge and skills that produce jobs and increase productivity. It equips people for meeting contemporary needs.

Recognizing that the countries as well as the regional and international organizations will develop their own priorities and schedules, it is essential to incorporate sustainable development concepts into all levels of education, from basic to tertiary and for all groups of society. This requires the development of new and alternative teaching methods and the strengthening of community involvement and educational partnerships. The potential for indigenous knowledge to contribute to educational efforts should not be overlooked. In addition, governments might, where necessary, help relevant NGOs promote their access to and involvement in the educational system.

A major priority is to reorient the education towards sustainable development by improving each country's capacity to address environment and development in its educational programmes, particularly in basic learning. This is indispensable for enabling people to adapt to the swiftly changing world and to develop an ethical awareness consistent with the sustainable use of natural resources. Education should, in all disciplines, address the dynamics of the physical/biological and socio-economic environment, and human development, including spiritual development should employ both formal and non-formal methods of communication.

Governments should strive to ensure, by the year 2000, a universal access to basic education and to reduce current adult illiteracy rates. They might prepare or update national strategies for environmental and developmental, education, both formal and informal, policies, and current activities. Schools should be assisted in signing environmental activity work plans, with the participation of students, staff, and incorporate them throughout the curriculum. They should employ rover and innovative interactive teaching methods.

Governments could support university activities and networks, and establish or strengthen national or regional, centres of excellence for interdisciplinary research and education on environmental and developmental issues. The UN system should undertake a comprehensive review of its educational activities and establish a programme within two years to integrate the decision and similarly that of the Rio Conference into the existing UNl educational framework.

Another major priority is the promotion of public awareness. There remains a considerable lack of awareness of the interrelatedness of all human activities and the environment, due to inaccurate or insufficient information. Developing countries, in particular, lack relevant technologies and expertise. Public sensitivity to environmental and developmental problems must be increased along with a sense of personal responsibility and greater motivation and commitment towards sustainable development.

Countries should strengthen and establish advisory bodies to act as catalysts for public environment and development information. They should promote a cooperative relationship with the media, entertainment and advertising industries and initiate discussions to mobilize their experiences in shaping the public behaviour and consumption patterns. In raising public awareness, modern communication technologist should be utilized to ensure that all sectors of society are being taught. Environment-related leisure and tourism activities should be supported, such as museums, zoos and national parks. Public awareness should be heightened regarding the impacts of violence in society.

The educational establishment, particularly, the tertiary sector, should be encouraged to contribute more to overall awareness building. NGO should increase their involvement through joint awareness initiatives and improve interchanges with other-sectors of society. Countries should increase their interaction with indigenous people and their communities and develop support programmes for women children and youth. The UN should improve its channels out of reach and interaction with governments in the above activities.

The UN-supported countries should identify workforce and tanning needs, and should integrate environmental and developmental issues into current training curricula. They should encourage all sectors of society, such as industries and universities, to include and environmental management component, in relevant training activities. Countries should also establish on practical training programmes for graduates from vocational schools, high schools and universities in order to enable them to meet labour market requirements and achieve sustainable livelihoods, together with industry trade unions and consumer groups. Governments should promote an understanding of the interrelationships between good environment and good business practices.

Sustainable development education requires the training of teachers, scientists, the business community, government officials and members of society on basic sustainable development concepts, based on up-to-date information. In building capacities and developing human resources, the identification of sector-specific training programmes is important.

For all the above educational activities, additional funding is essential, although much could be done through the reallocation of existing facilities, training arrangements and increasing community contributions.

REVIEW QUESTIONS

1. Define the term resources and natural resources.
2. What are renewable and non-renewable resources?
3. List out different natural resources.
4. What is deforestation? What are the causes of deforestation?
5. Discuss the ill effects of deforestation.
6. How are tribal people affected by deforestation?
7. With the help of a case study, explain the following with respect to environmental degradation. Timber extraction, mining activity.
8. Discuss the effect of dam construction on forests and tribal people.
9. What are the measures to be taken in the conservation of forests?
10. What is the importance of water in our life? How is water important for human development?
11. "World water conflict will be forever". Comment on this statement.
12. What are the causes of overutilization of ground and surface water? How can this lead to ill effects?
13. Discuss the benefits of dams and the problems in the construction of dams.
14. Explain with reference to human environment:
    i. Water crisis
    ii. Floods
    iii. Drought
15. What are the ill effects of mineral exploitation on the environment.
16. What are the uses of minerals? How does overexploitation affect the environment.
17. Explain an example of the ill effects of mineral exploitation.
18. Discuss the major food problems faced by the world.

19. Comment on whether India has sufficient food.

20. Discuss the impact of overgrazing.

21. What are the problems of modern agriculture?

22. What are the problems and causes of waterlogging?

23. What are the problems and causes of salinity?

24. How can the problem of salinity be rectified?

25. Classify and describe the energy resources.

26. Discuss the energy scenario in India.

27. Discuss the potential of non-conventional energy resources in India.

28. What are alternate energy sources? Why are they necessary?

29. How is solar energy utilized in India?

30. Discuss the importance of geothermal energy. How is geothermal energy useful in the generation of power.

31. Discuss bio-energy as a nonconventional energy. How is this energy utilized in the rural and urban areas of India?

32. How is tidal energy useful to mankind? What are its uses in India?

33. What are the demerits and merits of wave energy? What are the principles behind the Ocean Thermal Energy Convention?

34. Write notes on:
    i.   Wind energy
    ii. Wave energy
    iii. Tidal energy
    iv. OTEC

35. Explain land as a resource.

36. How is the degradation of land taking place? How can it be avoided?

37. Discuss the factors responsible for soil erosion? How can soil erosion be avoided?

38. What is desertification? What are the causes of desertification? How can desertification be avoided?

39. Explain the equitable use of resources for sustainable development.

40. How can we conserve natural resources?

CONCEPT OF AN ECOSYSTEM

Ecology deals with the interrelationship between the living organisms and their environment. Environment actually means the immediate surroundings of living organisms, which influence their lives either directly or indirectly (Figure 9.1).

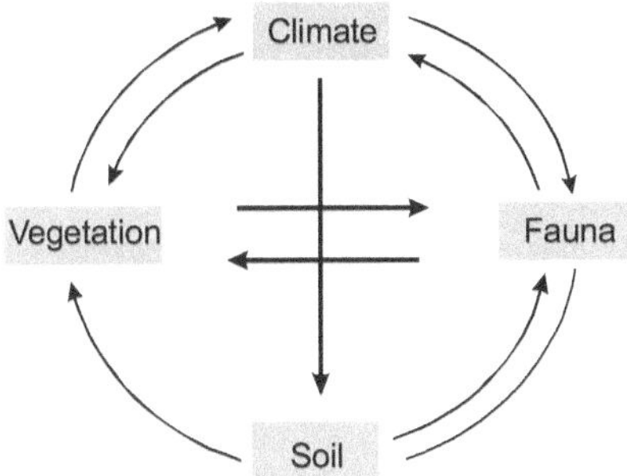

Figure 9.1 Reciprocal relationship in ecosystem

Ecology, a word derived from the Greek word *Oikos*, meaning house, has a wide scope. Broadly speaking, ecology is the study of the relationships of organisms to their environment and to other organisms.

An ecosystem consists of living matter (biotic) and non-living (abiotic) organic and inorganic matter such as water, gases, minerals, etc. A pond or a forest can also be an example of ecosystem.

An ecosystem has no particular size. In a sense, asking how big an ecosystem is would be like asking how big a city government, an engine, or a triangle is. All of these things are of abstractions. A pond is an ecosystem and so is a forest. No two ecosystems are exactly alike, just as no two city government is alike.

The ecosystem (Figure 9.2) is considered to be an open system as energy and matter continuously flow into and out of it.

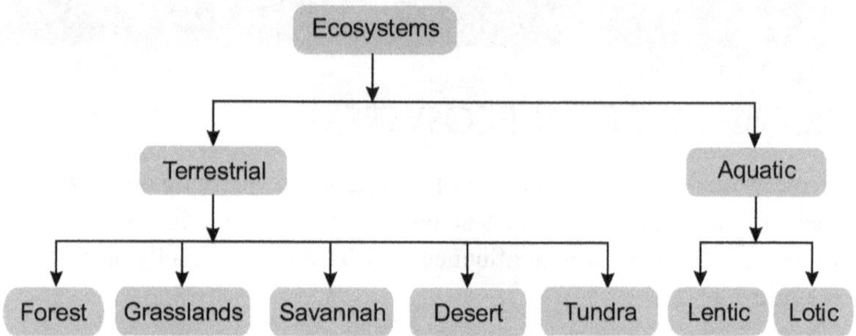

Figure 9.2   Ecosystems

STRUCTURE AND FUNCTIONS OF AN ECOSYSTEM

Structure   All ecosystems consist of abiotic (non-living) and biotic (living) constituents (Figure 9.3).

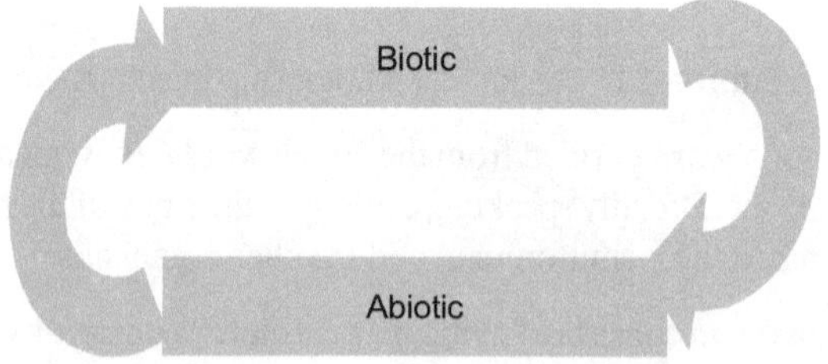

Figure 9.3   Components of an ecosystem

The flow of energy in an ecosystem occurs from one trophic level to the other. The producers convert the solar energy into chemical energy and store it as their energy. The consumers feed upon the producers and obtain energy, but they cannot accumulate all the energy since a major portion is lost in the form of heat energy. The energy flow in an ecosystem is always unidirectional, because the energy acquired from the sun can never be returned to the sun (Figure 9.4).

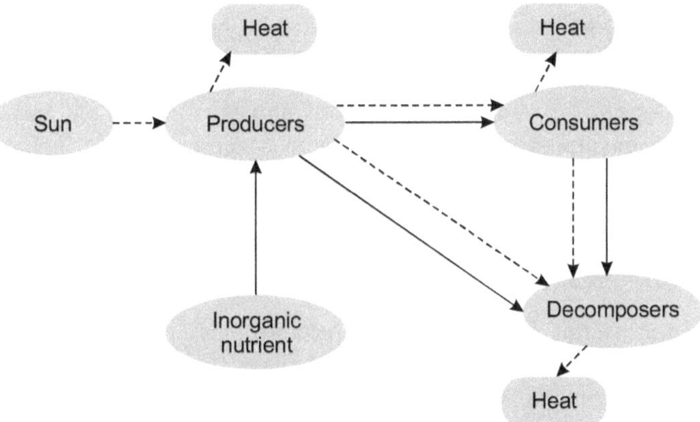

Figure 9.4   Nutrient cycle and energy flow

Abiotic Components

The abiotic components include the non-living components—soil, water, nutrients found in soil, climatic factors like temperature, rainfall, light and so on.

Biotic Components

The biotic components are the living components. In an ecosystem, there are three biotic components.

1.   Autotrophs
2.   Heterotrophs
3.   Decomposers

Autotrophs (Producers)   Autotrophs are producers ("Auto" means self and "troph" means nourishing). Autotrophs are green plants (trees, grasses, crops, tiny planktons, etc.). They possess a green pigment called chlorophyll. They are also called **photoautotrophs** because they utilize light energy. They convert solar energy into chemical energy with the help of inorganic substances like water and carbon dioxide and organic substances like enzymes. There are other types of autotrophs called **chemoautotrophs**, which use energy generated in an oxidation–reduction process. Examples of chemoautotrophs are sulphur bacteria, *Beggiatoa*.

Heterotrophs (Consumers)   Heterotrophic organisms ("*hetero*" means others, "*trophic*" means nourishing) are the consumers, chiefly the animals.

Depending upon their food habits, consumers may be **macroconsumers** or **microconsumers**. Macroconsumers include the **herbivores** which are plant eaters (e.g. insects, cattle), the **carnivores** which are flesh eaters (e.g. lion, tiger) and the **omnivores** which feed on both plants and animals (e.g. man, crow). Microconsumers include the **saprotrophs** or decomposers.

Decomposers   Decomposers are heterotrophic organisms but depend upon the dead organic matter for their food. They are chiefly microorganisms like bacteria and fungi. These break down complex matter like cellulose, etc. found in plant and animal body and ultimately release the simpler substances into the environment. Decomposers may be either **primary decomposers** which include microorganisms or **secondary decomposers** which include invertebrates.

Some of the important features of an ecosystem are listed below.

- Interaction between abiotic and biotic constituents is the main function of ecosystem.

- Sunlight is the ultimate energy source for primary interactions.

- Producers are plants. Plants use carbon dioxide and water to manufacture starch through photosynthetic action.

$$6CO_2 + 12H_2O \xrightarrow[\text{Sunlight}]{\text{Chlorophyll}} C_6H_{12}O_6 + 6O_2 + 6H_2O$$

- Producers use this starch as food source.

- The other function of an ecosystem is the distribution of energy in the form of food to consumers.

- Consumers are also fed by producers in ecosystem

◇ All living creatures should die one day or other and become abiotic components of the ecosystem, which are acted upon by decomposers.

◇ The third function of the ecosystem is decomposition, where the recycling of energy and materials are recycled. This recycling is called the **bio-geochemical cycle**.

GRASSLAND ECOSYSTEMS

Figure 9.5 Grassland ecosystems

Abiotic material The abiotic components of grasslands include the soil, nutrients found in the soil, organic substances like proteins, lipids and carbohydrates found in the dead organic matter which forms part of the soil, and climatic factors like temperature, rainfall, light, etc (Figure 9.5).

Producers   Producers are predominantly herbaceous plants which convert solar energy into chemical energy of food material (carbohydrates) with the help of chlorophyll, inorganic abiotic materials like carbon dioxide found in the air, and the macro- and micronutrients and water absorbed by roots from the soil in a biochemical process called **photosynthesis**.

Consumers   Insects, cattle, sheep and goat are the main consumers of grass and other herbaceous plants. Many invertebrates are 'below-ground' consumers. Several insects, rabbits, cattle, sheep, goat and many below-ground invertebrates are herbivores. The main carnivores are snakes, birds and hyaenas. Some insects and frogs are carnivorous (insects prey on other insects, frogs prey on insects, earthworms, etc).

Decomposers   Bacteria, actinomycetes, fungi, protozoa, and some other invertebrates present in soil act as decomposers. They degrade the dead organic matter of plant and animal origin and in the process, minerals are released into the soil, and again become available to producers. The decomposers are heterotrophic organisms (saprophytes) and are very diverse. They are found in the soil and on plant bodies.

POND ECOSYSTEM

The basic units in this ecosystem (Figure 9.6) are as follows:

I	Abiotic substances–basic inorganic and organic compounds
IIA	Producers–rotted vegetation
IIB	Producers–phytoplankton
III-IA	Primary consumers(herbivores)–bottom forms
III-IB	Primary consumers(herbivores)–zooplankton
III-2	Secondary consumers(carnivores)
III-3	Tertiary consumers (secondary carnivores)
IV	Saprotrophs–bacteria and fungi or decay

Abiotic materials   These include water, the nutrients found in water, and sediments and organic matter like proteins, lipids, carbohydrates, etc.

found in the water. Wind, water current, temperature, and the availability of sunlight are other abiotic factors (Figure 9.6).

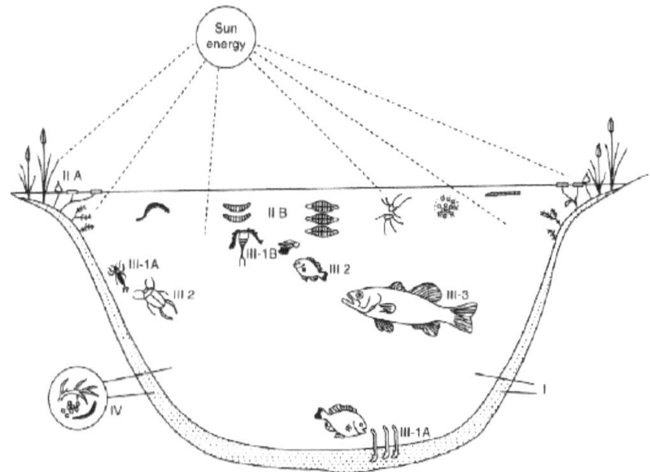

Figure 9.6   Pond ecosystem (*Source:* Odum, 1969)

Producers   Producers usually include microscopic phytoplankton which float in water and utilize the carbon dioxide and bicarbonates found in water. With the help of the chlorophyll present in their body, they convert solar energy into the chemical energy of food material.

Consumers   Zooplankton are the main herbivore consumers, which graze on phytoplankton. Some herbivorous fish also graze on phytoplankton. The carnivorous consumers are mainly small fish and some invertebrate animals. The top-order carnivores are the large fish.

Decomposers   Bacteria, actinomycetes, fungi and some invertebrate animals act as decomposers. They are found in water and in the sediments. They decompose dead organic matter and release the mineral substances from it. These minerals again become available to producers.

PRODUCTIVITY IN AN ECOSYSTEM

Productivity refers to the rate of production of organic matter on a unit area basis.

A certain portion of gross production is utilized by plants for maintenance and the remainder is called net primary production which is constituted by new plant biomass.

The rate at which the consumers convert the chemical energy of their food into their own biomass is called secondary productivity.

ENERGY FLOW IN ECOSYSTEM

In order to carry out a work, energy is required. Biological activity requires utilization of energy which ultimately comes from the sun. Solar energy is transferred into chemical energy by the process of photosynthesis and then transformed into mechanical and heat forms. In the biological world, the energy flows from the sun to plants and then to all heterotrophic organisms (Figure 9.7).

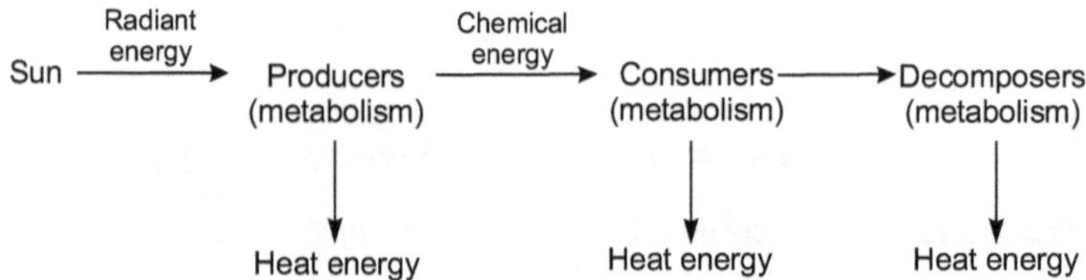

Figure 9.7   Energy flow in ecosystem

The flow of energy through an ecosystem can be explained in relation to the laws of thermodynamics which are applied to closed systems.

Laws of Thermodynamics

1. First law of thermodynamics states that energy can neither be created nor be destroyed.

2. Second law of thermodynamics states that if an increase or decrease occurs in the internal energy of the system itself, work is done and heat is either evolved or absorbed.

The decrease in the internal energy of the system	=	Work done by the system	+	Heat given off by the system

Thermodynamics in Ecosystems

In ecosystem solar energy is converted into chemical energy stored in food materials, which is converted into mechanical and heat energy. Hence, in ecological systems, energy is neither created nor destroyed but is converted from one form into another.

The quantity of solar energy that enters the earth's atmosphere is about 15.3×10^8 cal/m²/year. As much as 95–99% of the energy is lost from autotrophs in the form of heat of evaporation and sensible heat. The remaining 1–5% is used in photosynthesis for primary production.

Energy Flow at Each Trophic Level

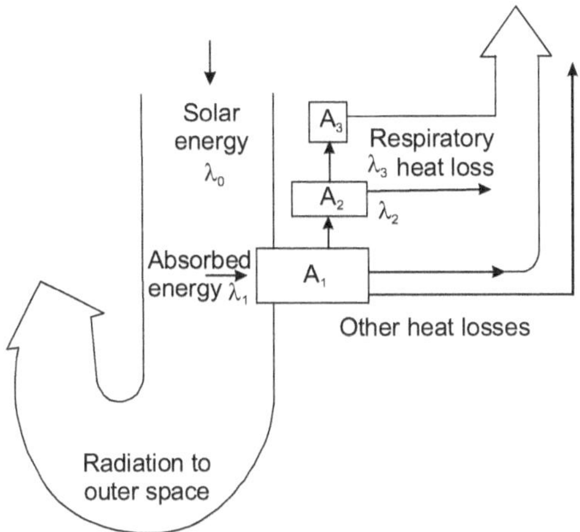

Figure 9.8   Energy flow at trophic level

Figure 9.8 shows that energy is transferred between trophic levels. Energy transfer is not 100% efficient and there is degradation of energy from one level to another as there is loss of energy as heat.

ECOLOGICAL SUCCESSION

The structure of an ecosystem changes constantly in response to changing environmental conditions. The gradual change in species composition of a given area is called ecological succession.

During succession, some species colonize an area and their population becomes more numerous whereas the population of other species declines and the species may even disappear.

Ecological succession is of two types:

1.   Primary succession
2.   Secondary succession

Primary Succession

Primary succession involves the gradual establishment of biotic communities on nearly lifeless ground (Figure 9.9).What was once base rock becomes a complex forest community.

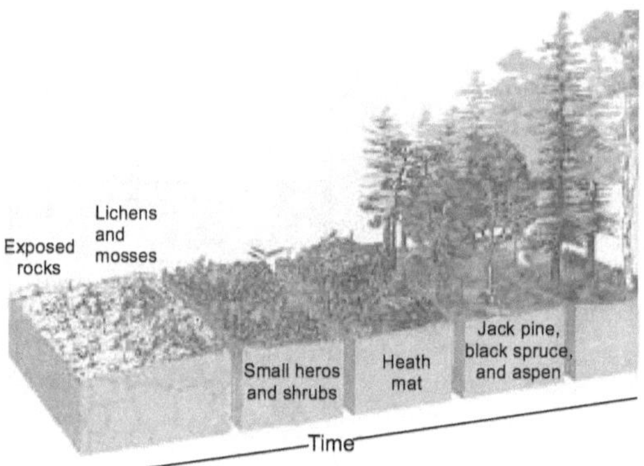

Figure 9.9   Primary ecological succession

Secondary Succession

The establishment of biotic communities in an area where some type of biotic community is already present is called secondary succession. This is the more common type of succession (Figure 9.10).

Secondary succession begins in an area where the natural community of organisms has been disturbed, removed or destroyed but some soil or bottom sediment remains. The area for secondary succession includes abandoned croplands, burned or cut forest, heavily polluted streams, and land that has been dammed or flooded.

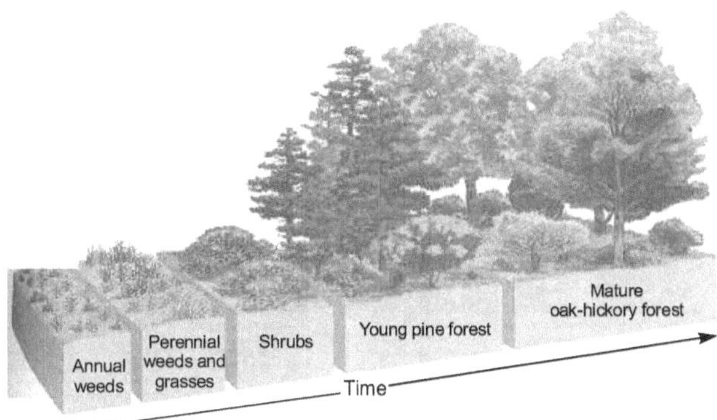

Figure 9.10   Secondary ecological succession

Importance of Ecological Succession

⬦   Describes the changes in vegetation.

⬦   These changes inturn affect food and shelter for various types of animals.

⬦   As succession proceeds, the number and type of animals and decomposers also change.

⬦   Many disturbances like fire, flooding, overgrazing and pesticide application can devastate communities and ecosystems.

Such disturbances create new conditions that can discourage or eliminate some species but encourage others by releasing nutrients.

FOOD CHAIN AND FOOD WEB

Plants or autotrophs convert solar energy into chemical energy that is stored as food materials. The transfer of food energy from plant sources through a series of organisms forms what is called a **food chain**.

A food chain observed in an Indian river is as follows:

phytoplankton ⟶ zooplankton ⟶ small fish ⟶ large fish ⟶ man

In an Indian pasture the following food chain is observed.

grass ⟶ grasshopper ⟶ toads ⟶ ratsnake

In these two examples, the base of the food chain is formed by a plant which is grazed on by a herbivore, which is predated over by a carnivore, which in turn may be eaten by another carnivore (Figure 9.11 and 9.12).

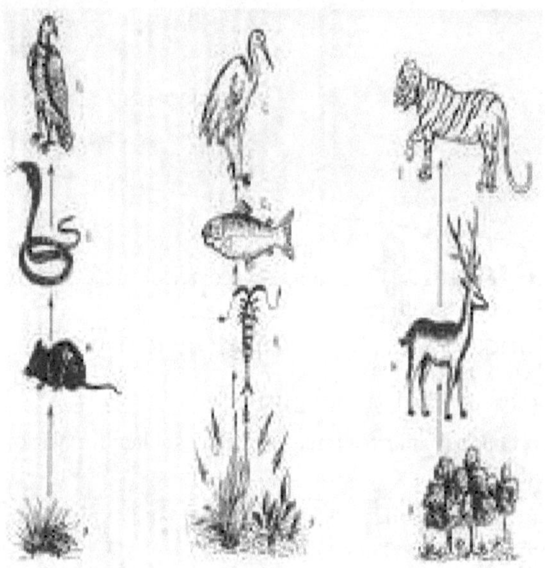

Figure 9.11   Food chains in nature: (a) Food chain in a grassland
(b) Food chain in a pond   (c) Food chain in a forest

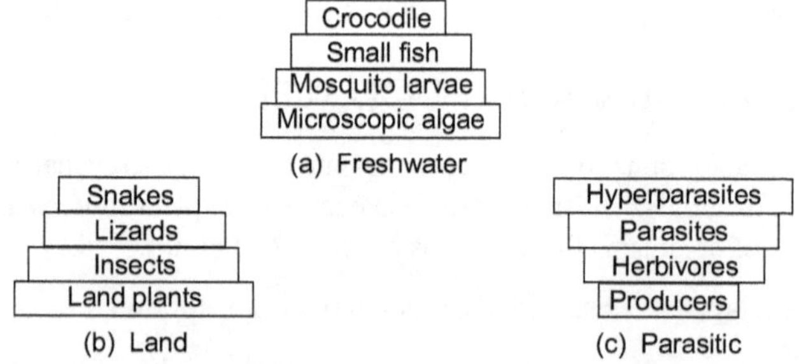

Figure 9.12   Examples of a food chain

Most of the organisms will depend upon more than one source of food. They consume a variety of food. For example, grass may be eaten by grasshoppers as well as rabbits or cattle and each of these herbivores may be eaten by many carnivores, such as toads, snakes, or birds depending on their food habits. This is called **food web**.

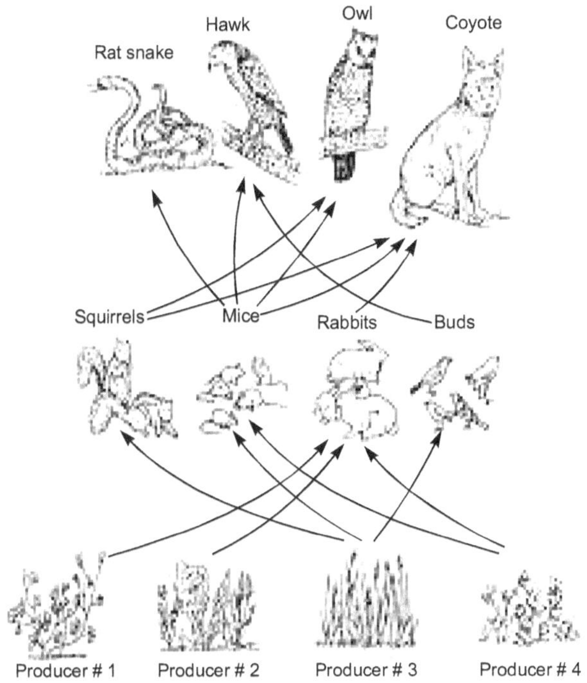

Figure 9.13   Food web

Significance of Food Chain

The food chain helps us to understand the feeding interrelationship between organisms in an ecosystem.

It also helps us in comprehending the energy flow mechanism and circulation of matter in the ecosystem and in understanding the movement of toxic substances which get biomagnified in our body since man is an omnivore and has access to all trophic levels for food.

There is a limit to the number of links in any food chain. The number of links very rarely exceeds five because in the process of energy transfer there is loss of energy to the environment in the form of heat.

Trophic Level

Organisms obtain their food from plants by one or more steps. Organisms whose food is obtained from plants by the same number of steps are said to belong to same trophic level

- ◇ Green plants occupy the first trophic level (producers level)

- ◇ Plant grazers belong to the second trophic level (primary consumers or herbivores)

- ◇ Flesh eaters eat herbivores and belong to the third trophic level— secondary consumers.

The next trophic level is tertiary consumer.

ECOLOGICAL PYRAMIDS

In any ecosystem the number of organisms at the base of the food chain is relatively much larger than that at the top. There is a progressive decrease in between the two extremes. This fact is called the **pyramid of numbers**. The concept of the ecological pyramid was developed by Charles Elton and so the name **Eltonian pyramids**.

There are three types of ecological pyramids now recognized.

- i. Pyramid of numbers
- ii. Pyramid of biomass
- iii. Pyramid of energy

These ecological pyramids provide a common denominator for the comparison of different communities.

Pyramid of Numbers

Carnivores sit at the top of food chain. The number of organisms gradually decreases from the base to the top of the pyramid in the forest or lake ecosystem. However, in a parasitic food chain the pyramid is always inverted. The number of organisms gradually increases making the pyramid inverted in shape (Figure 9.14).

Figure 9.14   A Pyramid of numbers

Pyramid of Biomass

The total dry weight of the total living material at any one time in a food chain is called the **biomass**. In a grassland or forest ecosystem there is gradual decrease in the biomass at successive levels, from the producers to the consumers. These pyramids are upright in position.

Figure 9.15 illustrates the pyramid of biomass showing the biomass production at various trophic levels in the world's oceans.

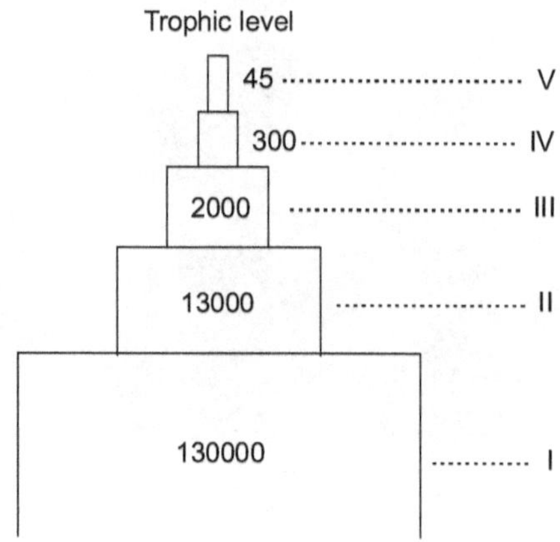

Figure 9.15   Pyramid of biomass

Pyramid of Energy

The pyramid of energy is a picture of the status of passage of food mass through the food chain. There is always a gradual decrease in the energy content at successive trophic levels from the producer to the consumer. The pyramid if energy is upright (Figure 9.16).

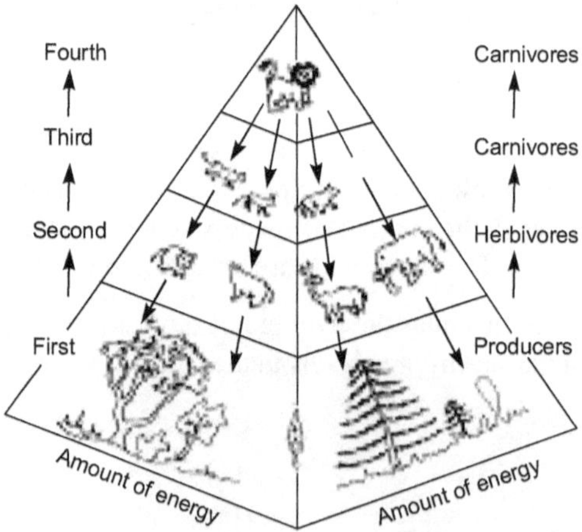

Figure 9.16   An ecological pyramid of energy

India is recognized to be uniquely rich in biodiversity. Here, almost all the biogeographic zones of the world are represented. According to a recent classification done by the Wildlife Institute of India, the country's biological wealth can be seen as representing about ten broad biogeographical zones (Table 2.1), namely Trans-Himalayan, Himalayan, desert, semi-arid, Western Ghats, Deccan peninsula, Gangetic plain, North-east India, islands, and coast. Each of these ten biogeographic zones has characteristic biota, and broadly represents similar climatic conditions and constitutes the habitat for diverse species of fauna and flora.

TRANS-HIMALAYAN ZONE

The Trans-Himalayan zone, with its sparse mountain vegetation type, has the richest wild sheep and goat community in the world. The snow leopard is found here, as is the migratory black-necked crane.

HIMALAYAN ZONE

The Himalayan zone extends from the north-west region of Kashmir to the east up to North Eastern Frontier (NEF). It comprises four biotic provinces—north-west, west, central and east Himalayas. Altitudinally there are three zones of vegetation in the Himalayan zone corresponding to the three climatic belts.

DESERT ZONE

The desert zone comprises three biotic provinces, viz. Kutch, Thar and Ladakh. The north-west desert region (Kutch and Thar) consists of parts of Gujarat, Rajasthan, Haryana and Delhi. The climate of this region is characterized by very hot and dry summer, and cold winter. Rainfall here is less than 700 mm.

SEMI-ARID ZONE

Adjoining the north-west desert are the semi-arid areas comprising Madhya Pradesh, Chattisgarh, parts of Orissa and Gujarat. Depending upon the amount of rainfall, the forests in this region have developed into thorny, scrub mixed deciduous and sal type.

WESTERN GHATS ZONE

The Western Ghats zone comprises the Malabar coast and Western Ghat mountains of India extending from Gujarat in the north to the Cape Comorin in the south. Rainfall in this region is heavy. The vegetation is of four types—tropical moist evergreen forests, subtropical or temperate evergreen forests, mixed deciduous forests and the mangrove forests.

DECCAN PENINSULA ZONE

The Deccan Peninsula zone comprises five biotic provinces, viz. Deccan Plateau (South), Central Plateau, Eastern Plateau, Chhota Nagpur Plateau and Central Highlands. The zone is a semi-arid region lying in the rain-shadow of the Western Ghats.

THE GANGETIC PLAIN

In the North is the Gangetic Plain extending up to Himalayan foothills. This region comprising Uttar Pradesh, Bihar and West Bengal is the most fertile region. The major climatic factors—temperature and rainfall—are together responsible for the distinctive type of vegetation in this zone. The rainfall varies from less than 700 mm in Western Uttar Pradesh to more than 1,500 mm in West Bengal. Vegetation is chiefly of the tropical moist and dry deciduous forest type.

NORTH-EAST INDIA

North-East India is one of the richest flora regions in the country. The region receives the heaviest rainfall, with Cherrapunji receiving as much as more than 10,000 mm of rainfall. The temperature and wetness are also very high, resulting in dense tropical evergreen forests.

THE ISLANDS

The Islands of Lakshadweep in the Arabian Sea, and Andaman and Nicobar Islands in the Bay of Bengal have a wide range of spreading coastal forests like mangroves and beech forests, and in the interior some of the best preserved evergreen forests of tall trees.

COAST

India has a coastline of about 7,516.5 km. Mangrove vegetation is the characteristic of estuarine tracks along the coast, for instance, at Pichavaram near Chennai and Ratna Giri in Maharashtra.

Table 10.1   India's major biogeographic zones

Biogeographic zone	Biotic province
Trans-Himalayan	Upper region
Himalayan	North-west Himalayas
	West Himalayas
	Central Himalayas
	East Himalayas
Desert	Kutch
	Thar
	Ladakh
Semi-arid	Central India
	Gujarat Rajwara
Western Ghats	Malabar coast
	Western Ghats

(Contd.)

Table 10.1   (Continued)

Biogeographic zone	Biotic province
Deccan Peninsula	Deccan Plateau (South)
	Central Plateau
	Eastern Plateau
	Chhota Nagpur Plateau
	Central Highlands
Gangetic Plain	Upper Gangetic Plain
	Lower Gangetic Plain
North-east India	Brahmaputra Valley
	North-eastern Hills
Islands	Lakshadweep
	Andaman
	Nicobar
Coast	West
	East

Carbon, hydrogen, oxygen, phosphorus, nitrogen, copper, water, iron are abiotic components of the environment. These abiotic components are retained in the reservoirs like air, water or land and are utilized by plants, i.e., are taken up by the living system. The energy moves from one trophic level to another in the food chain. Following their death, the living organisms are converted into minerals and the cycles are repeated.

Living organism after their death and decay are converted into minerals in soil which is in turn taken up by plants followed by animals which reaches the higher trophic levels of food chain. This continuous cycle of movement of nutrients is called biogeochemical cycle.

Macronutrients   Many elements or materials which are essential for life and are required in large quantities are called macronutrients. Macronutrients include carbon, nitrogen, oxygen, hydrogen, potassium, calcium, magnesium, sulphur and phosphorus.

Micronutrients   Some elements are required in small amounts and are called micronutrients. These are iron, manganese, copper, zinc, boron, molybdenum, vanadium, cobalt, chlorine, etc.

The biogeochemical cycles may be one of the following types:

1. Gaseous cycles like carbon, oxygen and nitrogen cycles
2. Water cycle or the hydrological cycle
3. Sedimentary cycles like sulphur and phosphorus cycles

An understanding of these various biogeochemical cycles is essential to understand the functioning of the ecosystem.

HYDROLOGICAL CYCLE

Water is the universal solvent and hence important for cycling of other matter in ecosystem. It is also essential for photosynthesis. Water is found in the liquid, solid and gaseous forms.

The hydrological or water cycle (Figure 11.1) is probably the most important of all natural cycles in the biosphere. The hydrological cycle refers collectively to the collection, purification and distribution of the earth's fiscal supply of water.

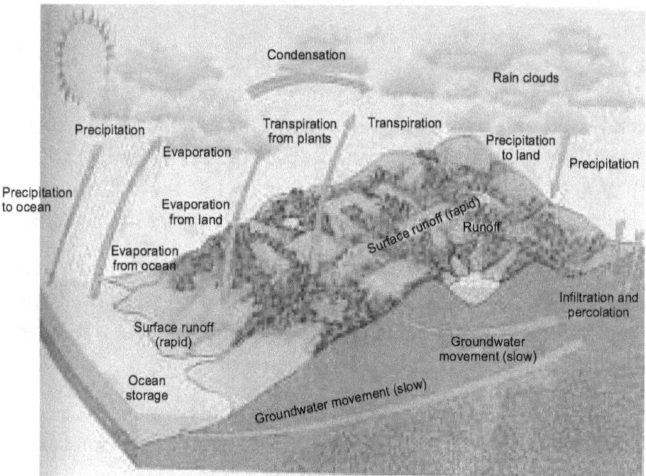

Figure 11.1   Simplified model of the hydrological cycle

Seawater under the hot sun evaporates, cools, condenses and precipitates under favourable conditions and maintains the water balance.

The main processes in the water cycle are

1. Evaporation—conversion of water into water vapour
2. Transpiration—evaporation from plant leaves after water is extracted from soil by roots and transported
3. Condensation—conversion of water vapour into droplets of liquid
4. Precipitation—rain, sleet, hail and snow
5. Infiltration—movement of water into soil

6. Percolation—downward flow of water through soil and rock strata

7. Run-off—surface movement of water down the slope back to the sea to resume the cycle

The water cycle is powered by energy of evaporation by the sun and by gravity. Incoming solar energy evaporates water from oceans, streams, lakes, soil and vegetation. About 84% of water vapour in the atmosphere comes from the oceans, the rest comes from land.

Some of the freshwater returning to the earth's surface by precipitation becomes locked in glaciers. Most of the precipitation falling on terrestrial ecosystems becomes surface run-off flowing into streams and lakes, which eventually carry water back to the oceans, where it can be evaporated to be cycled again.

Carbon Cycle

The carbon cycle is one of the biogeochemical cycles. The movement of carbon, in its many forms, between the biosphere, atmosphere, oceans, and geosphere is described by the carbon cycle. In the cycle there are various sinks, or stores, of carbon (represented by the boxes) and processes by which the various sinks exchange carbon (the arrows).

The carbon cycle is obviously very complex, and each process has an impact on the other processes. In the atmosphere, carbon exists almost entirely as gaseous carbon dioxide. The best estimates are that the earth's atmosphere contains 740 billion tons of this gas. Its global concentration is about 350 parts per million (ppm), or 0.035% by volume. That makes carbon dioxide the fourth most abundant gas in the atmosphere after nitrogen, oxygen and argon. Some carbon is also released as carbon monoxide to the atmosphere by natural and human mechanisms. This gas reacts readily with oxygen in the atmosphere, however, converting it to carbon dioxide.

Carbon returns to the hydrosphere when carbon dioxide dissolves in the oceans, as well as in lakes and other bodies of water (Figure 11.2). The solubility of carbon dioxide in water is not especially high, 88 millilitres of gas in 100 millilitres of water. Still, the earth's oceans are such a vast reservoir that experts estimate that approximately 36,000 billion tons of carbon are stored there. They also estimate that about 93 billion tons of carbon flows from the atmosphere into the hydrosphere each year.

Carbon moves out of the oceans in two ways. Some escapes as carbon dioxide from water solutions and returns to the atmosphere. That amount is estimated to be very nearly equal (90 billion tons) to the amount entering the oceans each year. A smaller quantity of carbon dioxide (about 40 billion tons) is incorporated into aquatic plants.

On land, green plants remove carbon dioxide from the air through the process of photosynthesis—a complex series of chemical reactions in which carbon dioxide is eventually converted to starch, cellulose, and other carbohydrates. About 10 billion tons of carbon are transferred to green plants each year, and a total of 560 billion tons of the element is thought to be stored in land plants alone.

The carbon in green plants is eventually converted into a large variety of organic (carbon-containing) compounds. When green plants are eaten by animals, carbohydrates and other organic compounds are used as raw materials for the manufacture of thousands of new organic substances. The total collection of complex organic compounds stored in all kinds of living organisms represents the reservoir of carbon in the earth's biosphere.

The cycling of carbon through the biosphere involves three major kinds of organisms. Producers are organisms with the ability to manufacture organic compounds such as sugars and starches from inorganic raw materials such as carbon dioxide and water. Green plants are the primary example of producing organisms. Consumers are organisms that obtain their carbon (that is, their food) from producers: all animals are consumers. Finally, decomposers are organisms such as bacteria and fungi that feed on the remains of dead plants and animals. They convert carbon compounds in these organisms to carbon dioxide and other products. The carbon dioxide is then returned to the atmosphere to continue its path through the carbon cycle (Figure 11.2).

Land plants return carbon dioxide to the atmosphere during the process of respiration. In addition, animals that eat green plants exhale carbon dioxide, contributing to the 50 billion tons of carbon released to the atmosphere by all forms of living organisms each year. Respiration and decomposition both represent, in the most general sense, a reverse of the process of photosynthesis. Complex organic compounds are oxidized with

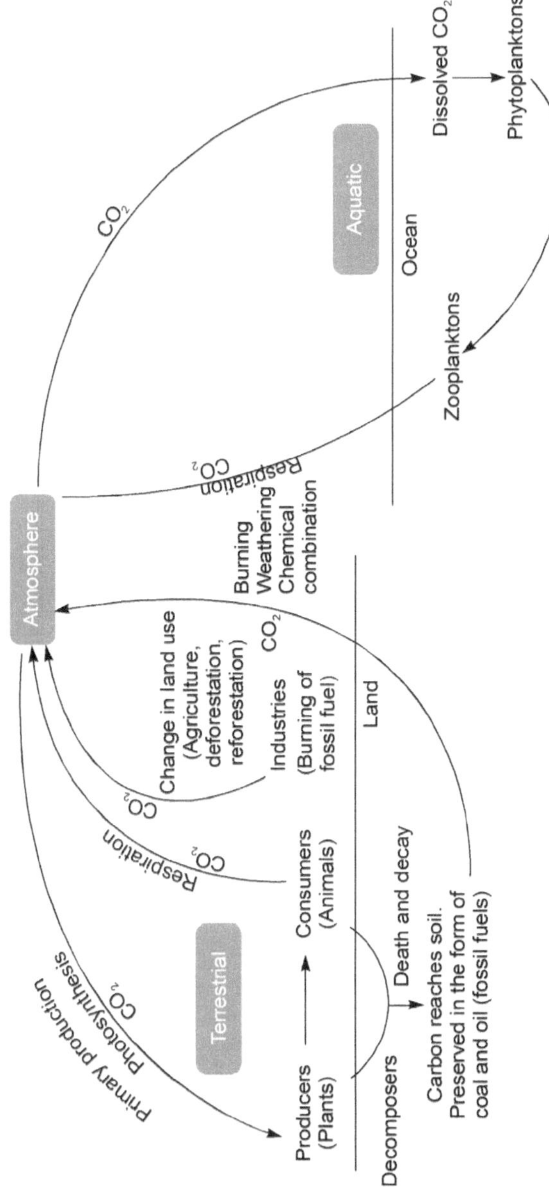

Figure 11.2   The carbon cycle

the release of carbon dioxide and water—the raw materials from which they were originally produced.

At some point, land and aquatic plants and animals die and decompose. When they do so, some carbon (about 50 billion tons) returns to the atmosphere as carbon dioxide. The rest remains buried in the earth (up to 1,50 billion tons) or on the ocean bottoms (about 3,000 billion tons). Several hundred million years ago, conditions of burial were such that organisms decayed to form products consisting almost entirely of carbon and hydrocarbons. Those materials exist today as pockets of the fossil fuels—coal, oil, and natural gas. Estimates of the carbon stored in fossil fuels range from 5,000 to 10,000 billion tons.

The processes that make up the carbon cycle have been occurring for millions of years, and for most of this time, the systems involved have been in equilibrium. The total amount of carbon dioxide entering the atmosphere from all sources has been approximately equal to the total amount dissolved in the oceans and removed by photosynthesis. However, a hundred years ago changes in human society began to unbalance the carbon cycle. The Industrial Revolution initiated an era in which the burning of fossil fuels became widespread. In a short amount of time, large amounts of carbon previously stored in the earth as coal, oil, and natural gas were burnt up, releasing vast quantities of carbon dioxide into the atmosphere.

Between 1900 and 1992, measured concentrations of carbon dioxide in the atmosphere increased from about 296 ppm to over 350 ppm. Scientists estimate that fossil fuel combustion now released about five billion tons of carbon dioxide into the atmosphere each year. In an equilibrium situation, that additional five billion tons would be absorbed by the oceans or used by green plants in photosynthesis. Yet this appears not to be happening: measurements indicate that about 60% of the carbon dioxide generated by fossil fuel combustion remains in the atmosphere.

Changing land use is a broad term which encompasses a host of essentially human activities. They include agriculture, deforestation, and reforestation. The problem is made even more complex because of deforestation. As large tracts of forest are cut down and burned, two effects result: carbon dioxide from forest fires is added to that from other sources, and the loss of trees decreases the worldwide rate of photosynthesis. Overall, it appears that these two factors have resulted in an additional one to two billion tons of carbon dioxide in the atmosphere each year.

No one can be certain about the environmental effects of this disruption of equilibria in the carbon cycle. Some authorities believe that the additional carbon dioxide will augment the earth's natural green house effect, resulting in long-term global warming and climate change. Others argue that we still do not know enough about the way oceans, clouds, and other factors affect climate to allow such predictions.

Nitrogen Cycle

Nitrogen cycle (Figure 11.3) is the biogeochemical cycle that describes the transformation of nitrogen and nitrogen-containing compounds in nature. It is a gaseous cycle.

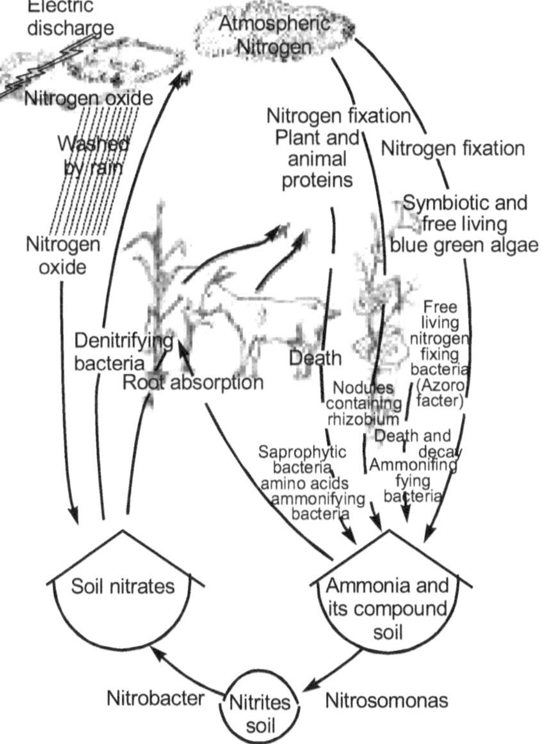

Figure 11.3   Nitrogen cycle

Earth atmosphere is about 78% nitrogen, making it the largest pool of nitrogen. Nitrogen is essential for many biological processes; and is crucial for any life on earth. It is present in all amino acids, is incorporated into proteins, and is present in the bases that make up nucleic acids, such as DNA and RNA. In plants, much of the nitrogen is used in chlorophyll molecules which are essential for photosynthesis and further growth.

Processing, or fixation, is necessary to convert gaseous nitrogen into forms usable by living organisms. Some fixation occurs in lightning strikes, but most fixation is done by free-living or symbiotic bacteria. These bacteria have the nitrogenase enzyme that combines gaseous nitrogen with hydrogen to produce ammonia, which is then further converted by the bacteria to make its own organic compounds. Some nitrogen-fixing bacteria, such as *Rhizobium*, live in the root nodules of legumes (such as peas or beans). Here they form a mutualistic relationship with the plant, producing ammonia in exchange for carbohydrates. Nutrient-poor soils can be planted with legumes to enrich them with nitrogen. A few other plants can form such symbioses. Nowadays, a very considerable portion of nitrogen is fixated in ammonia chemical plants.

Other plants get nitrogen from the soil by absorption at their roots in the form of either ions or ammonium ions. All nitrogen obtained by animals can be traced back to the eating of plants at some stage of the food chain.

Due to their very high solubility, nitrates can enter groundwater. Elevated nitrate in groundwater is a concern for drinking water use because nitrate can interfere with blood-oxygen levels in infants and cause methaemoglobinaemia or blue-baby syndrome. Where groundwater recharges stream flow, nitrate-enriched groundwater can contribute to eutrophication, a process leading to high algal, especially blue-green algal populations and the death of aquatic life due to excessive demand for oxygen. While not directly toxic to fish life like ammonia, nitrate can have indirect effects on fish if it contributes to this eutrophication. Nitrogen has contributed to severe eutrophication problems in some water bodies. As of 2006, the applications of nitrogen fertilizer is being increasingly controlled in Britain and the United States. This is occurring along the same lines as control of phosphorus fertilizer, restriction of which is normally considered essential to the recovery of eutrophied waterbodies.

Ammonia is highly toxic to fish life and the water discharge levels of ammonia from wastewater treatment plants must often be closely monitored. To prevent loss of fish, nitrification prior to discharge is often desirable. Land applications can be an attractive alternative to the mechanical aeration needed for nitrification.

During anaerobic (low oxygen) conditions, denitrification by bacteria occurs. This results in nitrates being converted to nitrogen gas and returned to the atmosphere. Nitrate can also be reduced to nitrite and subsequently combined with ammonium in the anammox process, which also results in the production of dinitrogen gas.

Nitrogen Fixation

Conversion of N_2   The conversion of nitrogen (N_2) from the atmosphere into a form readily available to plants and hence to animals and humans is an important step in the nitrogen cycle, that determines the supply of this essential nutrients (Figure 11.3). There are four ways to convert N_2 (atmospheric nitrogen gas) into more chemically reactive forms:

1. *Biological fixation*   Some symbiotic bacteria (most often associated with leguminous plants) and some free-living bacteria are able to fix nitrogen and assimilate it as organic nitrogen. An example of mutualistic nitrogen fixing bacteria are the *Rhizobium* bacteria, which live in legume root nodules. These species are diazotrophs. An example of the free-living bacteria is *Azotobacter*.

2. *Industrial N-fixation*   Haber–Bosch process, N_2 is converted together with hydrogen gas (H_2) into amino (NH_3) which is used to make fertilizer and explosives.

3. *Combustion of fossil fuels*   Automobile engines and thermal power plants, which release various nitrogen oxides (NO_3).

4. *Other processes*   Additionally, the formation of NO from N_2 and O_2 due to photons and especially lightening are important for atmospheric chemistry, but not for terrestrial or aquatic nitrogen turnover.

Assimilation   Plants can absorb nitrate or ammonium ions from the soil via their root hairs. If nitrate is absorbed, it is first reduced to nitrate ions and then ammonium ions for incorporation into amino acids, nucleic acids, and chlorophyll. In plants which have a mutualistic relationship with rhizobia, some nitrogen is assimilated in the form of ammonium ions directly from the nodules. Animals, fungi, and other heterotrophic organisms absorb nitrogen as amino acids, nucleotides and other small organic molecules.

Ammonification   When a plant or animals die, or an animal excretes, the initial form of nitrogen is organic. Bacteria, or in some cases, fungi, convert the organic nitrogen within the remains back into ammonia—a process called ammonification or mineralization.

Nitrification   The conversion of ammonia to nitrates is performed primarily by soil-living bacteria and other nitrifying bacteria. The primary stage of nitrification, the oxidation of ammonia (NH_3) is performed by bacteria such as the *Nitrosomonas* species, which convert ammonia to nitrates (NO_2). Other bacterial species, such as the *Nitrobacter*, are responsible for the oxidation of the nitrites into nitrates (NO_3).

Denitrification   Denitrification is the reduction of nitrates back into the largely inert nitrogen gas (N_2), completing the nitrogen cycle. This process is performed by bacterial species such as *Pseudomonas* and *Clostridium* in anaerobic conditions. They use the nitrate as an electron acceptor in the place of oxygen during respiration. These facultatively anaerobic bacteria can also live in aerobic conditions.

Anaerobic ammonium oxidation   In this biological process, nitrate and ammonium are converted directly into dinitrogen gas. This process makes up a major proportion of dinitrogen conversion in the oceans.

Human Influence on the Nitrogen Cycle

As a result of extensive cultivation of legumes (particularly soy, alfalfa and clover), growing use of the Haber–Bosch process in the creation of chemical fertilizers, and pollution emitted by vehicles and industrial plants, human beings have more than doubled the annual transfer of nitrogen into biologically available forms. In addition, humans have significantly contributed to the transfer of nitrogen trace gases from earth to the atmosphere, and from the land to aquatic systems.

N_2O has risen in the atmosphere as a result of agricultural fertilization, biomass burning, cattle and feedlots, and other industrial sources. N_2O has deleterious effects in the stratosphere, where it breaks down and acts as a catalyst in the destruction of atmospheric ozone. Ammonia (NH_3) in the atmosphere has tripled as the result of human activities. It is a reactant in the atmosphere, where it acts as an aerosol, decreasing air quality and clinging on to water droplets, eventually resulting in acid rain. Fossil fuel combustion has contributed to a 6- or 7-fold increase in NO_x flux to the atmosphere. NO_x actively alters atmospheric chemistry, and is a precursor of tropospheric (low atmosphere) ozone production, which contributes to smog, acid rain, and increases nitrogen inputs to ecosystems. Ecosystem processes can increase with nitrogen fertilization, but anthropogenic input can also result in nitrogen saturation, which weakens productivity and can kill plants. Decreases in biodiversity can also result if higher nitrogen availability increases nitrogen-demanding grasses thereby making the soil nitrogen-poor and causing a degradation of the species-diverse heathlands.

PHOSPHORUS CYCLE

The phosphorus cycle (Figure 11.4) is a biogeochemical cycle that describes the movement of phosphorus through the lithosphere, hydrosphere and biosphere. Unlike many other biogeochemical cycles, the atmosphere does

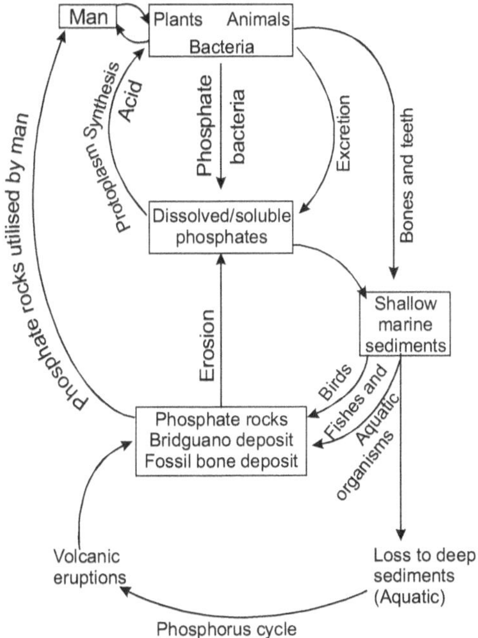

Figure 11.4   The phosphorus cycle

not play a significant role in the movements of phosphorus, because phosphorus and phosphorus-based compounds are usually solids at the typical ranges of temperature and pressure found on earth.

Phosphorus is an essential nutrient for plants and animals in the form of ions, PO_4^{3-} and HPO_4^{2}. It is a part of DNA molecules, of molecules that store energy (ATP and ADP) and of fats of cell membranes. Phosphorus is also a building block of certain parts of the human and animal body, such as the bones and teeth. The primary biological importance of phosphates is as a component of nucleotides, which serve for energy storage within cells (ATP) or when linked together, form DNA and RNA. Phosphorus is also found in bones, whose strength is derived from calcium phosphate, and in phospholipids (found in all biological membranes).

Phosphorus normally occurs in nature as part of a phosphate ion, consisting of a phosphorus atom and some number of oxygen atoms, the most abundant form (called orthophosphate) having four oxygens PO_4^{3-}. Most phosphates are found as salts in ocean sediments or in rocks. Over time, geologic processes can bring ocean sediments to land, and weathering will carry terrestrial. Plants absorb phosphates from the soil. The plants may then be consumed by herbivores which in turn may be consumed by carnivores. After death, the animal or plant decays, and the phosphates are returned to the soil. Run-off may carry them back to the ocean or they may be reincorporated into rock.

Phosphate move quickly through plants and animals; however, the processes that move them through the soil or ocean are very slow, making the phosphorus cycle overall one of the slowest biogeochemical cycles. However, recent findings suggest that phosphorus is cycled through the ocean on the timescale of 10,000 years, suggesting that the phosphorus cycle may play a role in global warming.

Unlike other cycles of matter compounds, phosphorus be found in air as a gas. This is because at normal temperature and circumstances, it is a liquid. It usually cycles through water, soil and sediments. In the atmosphere phosphorus is mainly small dust particles. Phosphorus is one of the longest cycles, and takes a long time to move from sediments to living organisms and back to sediments.

SULPHUR CYCLE

Sulphur cycle (Figure 11.5) is one of the constituents of many proteins, vitamins and hormones. It recycles as in other biogeochemical cycles.

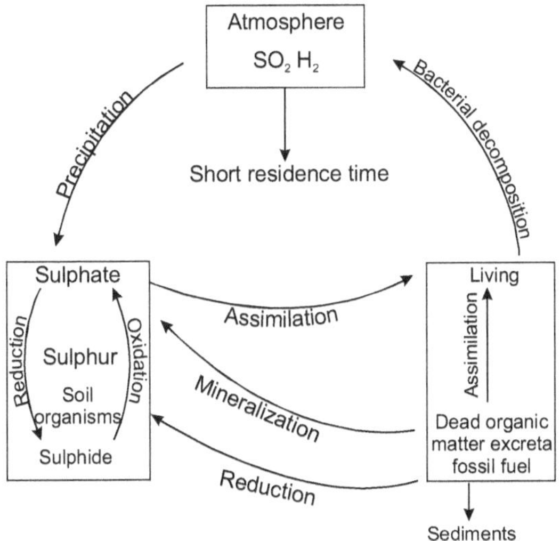

Figure 11.5   The sulphur cycle

The essential steps of the sulphur cycle are:

◇　Mineralization of organic sulphur to the inorganic form, hydrogen sulphide: (H_2S)

◇　Oxidation of sulphide and elemental sulphur (S) and related compounds to sulphate (SO_4^{2-})

◇　Reduction of sulphate to sulphide

◇　Microbial immobilization of the sulphur compounds and subsequent incorporation into the organic form of sulphur.

These are often termed as follows:

Assimilative sulphate reduction in which sulphate (SO_4^{2-}) is reduced to organic sulphydryl (otherwise known as thiol) groups (R–SH) by plants, fungi and various prokaryotes. The oxidation state of sulphur are +6 in sulphate and –2 in R–SH.

Desulphuration in which organic molecules containing sulphur can be desulphurated, producing hydrogen sulphide gas (H_2S), oxidation state = -2. Note the similarity to deamination.

Oxidation of hydrogen sulphide produces elemental sulphur (S^o), oxidation state = 0. This reaction is done by the photosynthetic green and purple sulphur bacteria and some chemolithotrophs.

Further *oxidation of elemental sulphur* by sulphur oxidizers produces sulphate.

Dissimilative sulphur reduction in which elemental sulphur can be reduced to hydrogen sulphide.

Dissimilative sulphate reduction in which sulphate reducers generate hydrogen sulphide from sulphate.

Human impact on the sulphur cycle is primarily in the production of sulphur dioxide (SO_2) from industry (e.g. burning coal) and the internal combustion engine. Sulphur dioxide can precipitate on to surfaces where it can be oxidized to sulphate in the soil (it is also toxic to some plants), reduced to sulphide in the atmosphere, or oxidized to sulphate in the atmosphere as sulphuric acid, a principal component of acid rain.

Our planet, earth, is occupied by diverse kinds of living organisms. They live in various environments. The world is estimated to have 5 to 30 million species of living organisms. At present about 2.5 million species of living organisms have been given scientific names. Over 1.5 million of them are animal species out of which 750,000 belong to insect species alone. There are 350,000 species of plants including algae, fungi, mosses and higher forms of plants.

The survival of such a vast range of living beings could be ensured only when their habitats and environmental conditions remain without alterations. The term "biosphere" had been coined to highlight the interdependence of the living and non-living world. It represents a stable environment in which various physical and biological factors which have been operating since the past. The organic continuity of the system rests on a delicate network of interdependent relationships. The air, the water, the animals, the plants, the microbes and human beings are all interlinked in a life sustaining system, called the environment.

Safeguarding the entire biosphere with all its intricacies is of prime importance today. The nations of the world have convened several conferences and adopted important resolutions for safeguarding the sustainability of earth. In this background, the United Nation's "Environmental Agency" organized the "International Conference on Human Environment" at Stockholm in 1972. This Conference adopted the motto 'Only one earth'. In 1982, a UN Conference on Environment was held at Nairobi. The UN again convened "Earth Summit" at Rio de Janeiro highlighting "Our Common Future", in 1992 once again a world summit on sustainable development was organized in Johannesberg in 2002.

One of the agenda commonly placed and accepted in all these meets was the significance of biodiversity and its conservation to ensure sustainable earth.

DEFINITION OF BIODIVERSITY

Biodiversity can be defined as species richness (plants, animals and microorganisms) occurring as an interacting system in a given habitat. It is the total sum of genes, species and ecosystems. Thus biodiversity conservation encompasses the whole spectrum of biota and their activities ranging from the macro-level of ecosystems to the micro-level of DNA libraries.

It can be generally described in terms of its three fundamental and hierarchically related levels of biological organization. These are:

1.   Genetic biodiversity
2.   Species biodiversity
3.   Ecological biodiversity

Genetic biodiversity   It is a measure of the variety available for the same genes within individual species.

Species diversity   It describes the different kinds of organisms within individual communities or ecosystems.

Ecological diversity   It is the richness and complexity of a biological community, including richness, number of trophic levels and ecological processes that capture energy, sustain food webs and recycle materials within this system.

FUNCTIONS OF BIODIVERSITY

There are two main functions of biodiversity. Firstly, the biosphere depends on biodiversity, which in turn leads to the stability in climate, water, soil, chemistry of air, and the overall health of the biosphere. Secondly, biodiversity is the source of species on which the human race depends for food, fodder, fuel, fibre, shelter, and medicine.

IMPORTANCE OF BIODIVERSITY

Biodiversity is not only an important resource but also the strength of the developing countries. Since biodiversity is an irreplaceable resource, and its extinction is forever, the conservation and sustainable utilization of this resource have to be central to all developmental planning in the developing countries. Economy of these countries is largely dependent on agriculture, horticulture, animal husbandry, fisheries, forestry, and medicinal and biodiversity has been kept alive by farmers, tribals and individual breeders.

The genes from wild ancestors of crops, endemic to the developing countries, have made significant contribution in crop improvement, which in turn has resulted into social and economic development.

Because of these reasons many organizations like the World Wide Fund for Nature (WWF) and the United Nations Commission on Environment and Development (UNCED) stressed the need to preserve our biodiversity.

BIOGEOGRAPHY

No organism occurs in uniform numbers throughout the world. Rather, specific organisms are restricted to specific communities or groups of communities. Three aspects of the distribution of an organism are generally recognized: the **geographic range**, or the specific extent of land or water area where the organism normally occurs; the **geologic range**, or the distribution in time, past and present; and the **ecological distribution**, or the major **biotic community** (i.e., marine biome, freshwater biome and terrestrial biome) of which the organism is a member. Certain biologists have also made distinction between **geographic distribution** (horizontal or superficial distribution) and **bathymetric distribution** (vertical or altitudinal distribution). Bathymetric distribution includes the following three realms: (i) **Halobiotic** or vertical distribution of organisms in marine (sea) habitat, (ii) **Limnobiotic** or vertical distribution of organisms in fresh water habitat; and (iii) **Geobiotic** or altitudinal distribution of organisms on land. All living organisms in a given region are termed the **biota** of that region. The animals of a given region are collectively termed the **fauna**, and the plants of a given region, the **flora** (i.e., fauna + flora = biota). Here, a distinction can be made between the two terms—flora and vegetation. Flora mainly refers to the botanical composition of a place, i.e., the names of different plant species, while vegetation means the totality of forms in which the emphasis is not on names of different plants but their life forms, number and coverage. The studies of the distribution of biota are collectively called **biogeography**; of animals only, **zoogeography**; and of plants only, **plant geography** or **phytogeography**. There are two major approaches to the study of biogeography or geographical ecology: (i) **descriptive or static biogeography** which deals with the description of biota of different botanical and zoological areas of the earth; and (ii) **interpretative or dynamic biogeography** which describes the forces which have brought about plant and animal distribution.

PHYTOGEOGRAPHY OF INDIA

The Indian subcontinent lying between 80° and 37° N, and 68° and 97° E has its own peculiar physiographic, climatic and biotic features. It is surrounded on its south, east and west by oceans and in the north by some of the highest mountain chains in the world. The subcontinent stretches out between tropical and subtropical belts. Its climate is chiefly modified by oceans and mountains. In the south and in the far east the climate is typically tropical, while it is temperate in the north, and highly arid in the north-west. The general climate of India is of the monsoon type. In India the following three distinct seasons occur:

1. Cold season during December to February when the mean temperature falls very low. In the north snowfall occurs on several days and the temperature remains below 0°C for many weeks, while in the south the temperature may remain above 20°C. Some rainfall occurs during this season.

2. From March to mid-June, the cold-weather season is followed by the hot dry summer as the temperature gradually rises, and relative humidity decreases. The change of season is marked by leaf-fall and blooming of many trees particularly in northern India. The summer in the north often becomes quite severe with average temperatures above 40°C or even up to 50°C in the arid western region. Hot dusty winds commonly blow at high velocity during this season.

3. Towards the end of the summer by late-June, the monsoon winds bring rain and the temperature gradually falls down. The rainfall starts little earlier in the eastern and southern parts of India. The monsoon winds moving from the Bay of Bengal up to the Gangetic plains of north to the west bring more rains in the east (up to 200 cm) and much lesser in the west (100 cm), till in the westernmost part

of the country, less than 50 cm rainfall is left. By the end of September the monsoon starts retreating, the sky becomes clear and the temperature gradually falls down up to December. This variability in the temperature and rainfall pattern produces a great diversity in the vegetation of the subcontinent.

VALUE OF BIODIVERSITY

The value of the earth's biological resources or biodiversity can be broadly classified into the following two categories.

1. Direct values
2. Indirect values

Direct Values

Direct values may be classified under four headings which include:

1. Food
2. Medicines
3. Raw materials for industries
4. Ethical and aesthetic values

Medicines

◇ In traditional medical practices like ayurveda, plants or their extracts are directly consumed or applied as medicines. In modern medicine too, around 199 pure chemical extracts from 90 species of plants are used in medicines.

◇ A host of microbial, antiviral, cardioactive and neurophysiological substances have been derived from marine fauna.

◇ Domesticated animals have given us hormones, enzymes, and food products while the fungi and microbes provide life-saving drugs such as antibiotics. The possibilities of medicines from biological resources are immense and the scope also is wide.

◇ Genes from a "wild wheat" in Turkey saved an epidemic of the wheat disease in USA in the 1960s. It provided resistance to this and 50 other diseases and is saving worth $ 50 million annually to the US alone.

◇   One "gene" from a single Ethiopian barley plant now protects California's $ 160 million annual barley crop from yellow dwarf virus.

◇   An ancient wild relative of corn (*Zea diploperennis*) from Mexico can be crossed with modern corn (Zea mays) varieties to produce a perennial high-yielding disease resistant variety of maize called *Zea perennis* with potential worldwide savings to farmers estimated at $ 4.4 billion annually in terms of agro-chemicals.

◇   In 1960, a child suffering from leukaemia had only one chance in five of survival. Now the child has four chances in five, due to the treatment with drugs containing active substances discovered in the rosy periwinkle, *(Catharanthus roseus)* a tropical forest plant originating from Madagascar.

◇   In 1970 when an epidemic called "grassy stunt virus" destroyed more than 1,60,000 hectares of rice in Asia, it could be controlled from a "single gene" of wild rice *Oryza nivara* from the forests of central India.

◇   A wild coffee gene from Ethiopia's fast dwindling forest saved the Brazilian coffee species in 1970.

Life-saving medicines from biodiversity   Worldwide, medicinal products from the wild, worth some $ 40 billion a year, are being utilized by humans. WHO has listed over 21,000 plant species, which have medicinal use.

Plants and other organisms are natural biochemical factories. More than 60 per cent of the world's people depend directly on plants for their medicines. The Chinese use more than 5,000 of the estimated 30,000 species of plants in their country for medicinal purposes.

Raw materials for industries   Wild plant-derived industrial materials include fats, oils, waxes, latexes, pentins, resins, gums and other exudates, vegetable dyes and tannins, lignin, cellulose, starch, hydrocarbons and a host of biochemical compounds.

Raw materials for explosive industries   Candelilla wax, which has great industrial potential, has been found suitable for end-use in the manufacture of explosives by the Cordite Factory, Nilgiris.

Raw materials for the drug and pharmaceutical industries   The root powder of *Withania somnifera* have been found to be "anabolic". It accelerates growth in children and retards the process of ageing in older people. Source of "biopesticides" are guayale (*Parthenium argentatum*) which grows wildly in the Thar desert of Rajasthan.

Raw materials for perfumeries   The essential oil extracted from the leaves and inflorescence of *Cymbopogon martinii* var. *motia* has been found to have great perfumery properties.

Food   The biodiversity provides food, medicines and clothes.

In future civilization has to depend upon "alternative foods" grown on a medium other than soil with a "short harvest cycle".

"Spirulina" occupies special significance among algal food and promises to revolutionize the food industry in world.

Discovery of some wild species of plant of potential economic value to mankind:

1.   Buffalo gourd (*Cucurbita foetidissima*)
2.   Amaranth (*Amaranthus polygamous*)
3.   Winged beans (*Psophocarpus tetragonolobus*)
4.   Crambe (*Crambe abyssinica*)
5.   Kenaf (*Hibiscus cannabinus*)
6.   Guayule (*Parthenium argentatum*)

Ethical and aesthetic values   Biodiversity teaches ethics from different perspectives.

There is an increasing awareness that humans have some kind of ethical responsibility for the welfare, or at least the continued existence, of our only known living companions like animals, birds and plants in the universe.

Aesthetic   A profound human instinct causes people to feel linked to the natural world. Even the most hardened city dwellers need space and greenery in their work and play.

The culture of every people is closely allied to its landscapes and their living inhabitants, and cannot be dissociated from them.

Biodiversity the variety of plants, animals and the microorganisms interacting with the surroundings and among one another are not only the "bio-indicators" of a healthy and sustainable human environment but also helps in the maintenance of a conducive environment essential for human survival.

Indirect Values

At the other end of spectrum, biological resources provide values which are not immediately seen but have far-reaching impact on our living conditions. Indirect benefits of biodiversity can be enumerated as follows:

- Carbon fixation through photosynthesis
- Pollination, gene flow, etc.
- Maintaining water cycles, recharging ground water, protecting watersheds and buffering from extreme conditions such as flood and drought
- Soil formation and protection from erosion
- Maintaining essential nutrient cycles
- Absorbing and decomposing pollutants
- Regulating climate at both macro- and micro-level
- Preserving recreational, aesthetic, socio-cultural, scientific educational, ethical and historical value of natural environments.

ECOLOGICAL SERVICES OF BIODIVERSITY

Man depends on natural ecosystems for essential ecological services like maintenance of atmosphere, cycling of nutrients, soil fertility, control of past outbreaks, maintenance of genetic library, etc.

i.   Species plays a small but significant role in the maintenance of whole ecosystem.

ii. Most species are superfluous. The biomass of primary producers, consumers and decomposers is important.

Only few species are needed to keep the system moving. Species richness increases productive efficiency of the system. Reduced biodiversity may alter the function of other ecosystems. Soil fertility is maintained and preserved by biodiversity. Addition or deletion of species provide direct evidence of the role of biodiversity in maintenance of ecosystems.

Removal of one species frequently leads to changes in the abundance and performance of other species. The impact may be so great that the very nature of the ecosystem may change. Extincting of even one species increases the probability of affecting the food chain followed by the damage to the whole ecosystem.

BIODIVERSITY AT NATIONAL AND LOCAL LEVELS

BIODIVERSITY AT THE GLOBAL LEVEL

Biodiversity is not distributed uniformly across the globe. It is substantially greater in some areas than the others. Generally, species diversity increases from the poles towards the tropics. Tropical moist forests cover only 5–7 per cent of the land but possess 50 per cent of the world species. On the other hand, certain regions though may not have a high diversity, display a high degree of endemism. However, a less diverse system does not mean that the region is not important (Table 12.1).

Table 12.1 Known and estimated diversity of life on earth

Form of life	Known species	Estimate number of species
Insects and other arthropods	874,161	30 million species, extrapolated from surveys in forest canopy in Panama; most believed unique to tropical forests
Higher plants	248,400	Estimates range from 275,000 to 400,000; at least 10–15 per cent species believed undiscovered
Invertebrates (excludes arthropods)	116,873	True invertebrates may number millions of species. Nematodes, eelworms, and roundworms may each comprise more than one million species
Lower plants (fungi and algae)	73,900	Not available
Microorganisms	36,600	Not available
Fish	19,056	21,000, assuming that 10 per cent fish remain undiscovered; the Amazon and Orinoco rivers alone may account for 2,000 additional species
Birds	9,040	Known species probably account for 98 per cent of all birds
Reptiles and amphibians	8,962	Known species probably account for over 95 per cent of all reptiles and amphibians
Mammals	4,000	Known species account for over 95 per cent of all mammals
Total	1,390,992	10 million species is considered a conservative estimate. If insect estimates are accurate, total exceeds 30 million

BIODIVERSITY IN INDIA

Biodiversity is indeed one of India's important strengths and is the bedrock of all bio-industrial developments in the unusually large rural sector (576,000 villages with 76 per cent of country's population) of the country.

India has over 108,276 species of bacteria, fungi, plants and animals already identified and described. Out of these, 84 per cent species constitute fungi (21.2 per cent), flowering plants (13.9 per cent) and insects (49.3 per cent). In terms of the number of species, the insects alone constitute nearly half of the biodiversity in India (Table 12.2).

These species occur on land, fresh and marine waters, or occur as symbionts in mutualistic or parasitic state with other organisms. In the world as a whole, 1,604,000 species of Monera, Protista, Fungi, Plantae and Animalia have been described so far. However, it is estimated that at least 1,7980,000 species exist in the world, but as a working figure 12,250,000 species are considered to be near reality.

Table 12.2   Number of species of bacteria, fungi, plants and animals in India

Taxon	Number of species	Percentage
Bacteria	850	0.8
Fungi	23,000	21.2
Algae	2,500	2.3
Bryophyta	2,564	2.4
Pteridophyta	1,022	0.9
Gymnosperms	64	0.1
Angiosperms	15,000	13.9
Insecta	53,430	49.3
Mollusca	5,050	4.7
Pisces	2,546	2.4
Amphibia	204	0.2
Reptilia	446	0.4
Aves	1,228	1.1
Mammalia	372	0.3
Total	**108,276**	**100**

Two regions from India namely the Eastern Himalayas and the Western Ghats are included as hot spots.

Broadly, the botanical hot spots lie in the (i) Western Ghats, (ii) North east India, (iii) Himalayas and (iv) Andaman and Nicobar Islands.

The Western Ghats particularly the southern Western Ghats known as Malabar is the major genetic estate with an enormous biodiversity of ancient lineage. Out of five locations identified by IUCN as threatened, three areas—Periyar Agastyamalai Hills, Silent Valley and Periyar National Park—are found in this region.

North east India represents the transition zone between the Indian, Indo-Malayan, Indo-Chinese biogeographic regions as well as meeting place of Himalayan mountains with that of peninsular India.

Some of the hot spots in this zone are Namdapha, Tirap, Sirohi, Dzuko, Tura, Buxa valley, Senchal and Tale Valley.

Western Himalayas is quite remarkable for its alpine flora. Gymnosperms are quite abundant. About 5000 flowering plants occur here, of which 800 species are endemic.

Some major hot spots of this region are Valley of Flowers, Pithoragarh, Gori Valley, Mandal–Chopta Valley, Karakorum and Ladakh.

The Andaman and Nicobar Islands have a rich flora under littoral and inland types. About 2500 species of flowering plants are found here of which 250 species are endemic.

Some of the hot spots of this region are North Andaman Islands, South reef Island, Spike Island and the little and Great Nicobars.

Sacred Forests

These are the real hot spots of biodiversity. They are protected by tribals due to some religious sanctity attached to them. These sacred forests contain a number of rare, endangered, endemic or interesting biota of the country.

They are known as "Devaskadu" in Karnataka, 'Devarahati' in Maharashtra, and "La-Kyntok" in Meghalaya.

Mangroves

Mangroves are the salt-tolerant forest ecosystems found mainly in the tropical and subtropical inter-tidal regions. The total area of mangroves in India is estimated to be 6740 km², of which the mangroves of Sunderbans itself contribute 4200 km², next being the Andaman and Nicobar Islands. The quality and quantity of mangroves in India have greatly declined. The following mangrove areas must be considered as hot spots for conservation purposes.

Andaman   North Andaman, Prolobjig, Bomlungta, Havelock Island and Wrafter Creek (South Andaman), Nicobar, Komorta, Noncowry Islands, Bolloo Chengappa Bay and a few sheltered bays on Great Nicobar Islands.

East coast   Sunderbans, Bhitarkanika Mangroves and Godavari and Krishna Deltaic Mangroves, Coringa Bay.

West coast   The Gulf of Kutch and Gulf of Khambhat, Vembanad in Kerala.

Wetlands and Swamps

India has rich variety of wetlands. Some major wetlands of India are Kolleru (AP), Wullar (J&K), Chilka (Orissa), Loktak (Manipur), Bhoj (MP), Sambhar (Rajasthan), Pichola (Raj) , Ashtamudi (Kerala), Harike (Punjab), Ujni (Maharashtra), Sukhna (Chandigarh), Renuka (HP), Kadar (Bihar), Nalsarovar (Guj) and Kanjli (Punjab).

INDIA'S RANK IN BIODIVERSITY

India is the 10th among the plant-rich countries of the world, fourth among the Asian countries, eleventh according to the number of endemic species of higher vertebrates (amphibia, birds and mammals), and tenth in the world as far as richness in mammals is concerned. Out of the 18 "Hot spots" identified in the world, India has four. These are Eastern Himalayas, North east India, Western Ghats and Andaman and Nicobar Islands. The two areas, Eastern Himalayas and Western Ghats, contain 5,332 endemic species of higher plants, mammals, reptiles, amphibia and butterflies, to name a few. The following crops which first grew in India and spread throughout the world include rice, sugar cane, Asiatic vignas, jute, mango, citrus, banana, several species of millets, spices, medicinals, aromatics, and ornamentals. India ranks sixth among the centres of diversity and origin as far as agro-biodiversity is concerned.

MEGA BIODIVERSITY CENTRES

Using the criteria of species richness the concept of mega diversity centres has been introduced. Following this concept, 12 countries/regions namely Mexico, Columbia, Ecuador, Peru, Brazil, Zaire, Madagascar, China, India, Malaysia, Indonesia and Australia have been classified as mega biodiversity centres (Table 12.3). At the global level 10 hot spots of tropical forests are identified to which later 8 areas are added. These areas collectively include 49,955 endemic species which constitute 20 per cent of the world's total plant species.

Table 12.3 Biodiversity in world

Region	Number of plants
Cape region South Africa	6000
Upland Western Amazonia	5000
Madagascar	4900
Philippines	3700
Borneo	3500
Eastern Himalayas (India)	3500
SW Australia	2830
Western Ecuador	2500
Columbian Choco	2500
Peninsular Malaysia	2400
Californian Floristic Province	2140
Western Ghats (India)	1600
Central Chile	1450
New Caledonia	535
Eastern Arc Mts (Tanzania)	535
SW Sri Lanka	500
SW Coted Tvorie	200

HOT SPOTS OF BIODIVERSITY

All those areas that support rich biodiversity, because of geologic formations and endemic flora and fauna, and exhibit exceptional scientific interest are called as hot spots. They contain outstanding examples of evolutionary processes of speciation and extinction. These are the areas that are severely threatened by human activities.

Hot Spots in India

Silent Valley (Kerala)   Massive destruction of pristine biodiversity due to the proposed hydroelectric dam in the 1970s. Several species are threatened with extinction.

Pooyamkutty Valley (Kerala)   Rich biodiversity threatened due to the construction of yet another dam.

Palni Hills (Tamil Nadu)   Destruction of pristine "Shola Forest" due to population pressure and invasion of tourists.

Nilgiri Hills (Tamil Nadu)   Original climax forest is being replaced by the tea and coffee plantations. Rapid urbanization and massive tourist invasion is also destroying large patches of vegetarian.

Himalayas (India/Nepal)   Massive biodiversity destruction due to population pressure, construction of hydroelectric dams, expansion of apple orchard (horticulture) and pine plantations for resin tapping.

Shivalik Hills (Himachal Pradesh)   Massive destruction due to mining activities.

Doon Valley (Rajasthan)   Massive deforestation due to mining activities.

Aravalli Hills (Rajasthan)   Massive habitat destruction due to mining activities and population pressure.

Chilka Lake (Orissa)   Destruction of rich aquatic biodiversity due to commercial exploitation for fish and shrimp farming.

Pulikat Salt water lake (Tamil Nadu)   Destruction of aquatic life due to siltation.

Narayan Sarovar (Gujarat)   Pristine wetland ecosystem threatened with destruction due to revived mining activities in the area. It has been denotified by the government of Gujarat.

Gir Forest (Gujarat)   Only home for the last surviving Asiatic lions, threatened with destruction due to population pressure.

Manas (Assam)   Ecological "Gold mine" of India rich in wildlife diversity, threatened with destruction due to population pressure and illegal hunting, poaching and wildlife trade.

Sariska (Rajasthan)   Tiger Reserve of India threatened with destruction due to mining activities for procuring marbles. Supreme Court of India stopped mining in 1993.

Loktak Lake (Manipur)   Pristine wetland and only home for the last surviving brow antlered deer. Lake threatened with drying up.

Bharatpur Lake and wetland (Rajasthan)   Habitat of wide variety and diverse species of birds and winter home for the Siberian Cranes. Threatened due to tourist invasion and urban expansion.

Thar Desert (Rajasthan)   Highly generic desert in world and seat of some elusive flora and fauna. Endemic species are threatened due to desertification intensified by population pressure of humans and expanding livestock.

Sunderbans (West Bengal)   Seat of "pristine mangrove forest" and biggest home for the tigers. Threatened due to erosion of coastal habitats.

Little Rann of Kutch (Gujarat)   Only home for the last surviving wild ass threatened with destruction.

Renuka Lake and wetland (Haryana)   Aquatic life threatened due to massive tourist invasion.

Andaman and Nicobar Island (India)   Elusive flora and fauna have become endangered due to tourist invasion.

Western Ghat (India)   Seat of richest biodiversity in India threatened with destruction due to intense development activities. The Konkan Railway Project has particularly threatened the flora and fauna of the region.

North Eastern Himalayas (India)   Richest biodiversity after the Western Ghat threatened due to population pressure and short cycled shifting agriculture practiced by the tribal inhabitations.

Dal lake (Jammu and Kashmir)   Lake is dying due to destruction of aquatic life and human enchroachment. The lake area, 59 sq.km in 1856 has shrunk to mere 24 sq.km in 2005.

Kodai Lake (Tamil Nadu)   Lake is dying due to destruction of aquatic life by vehicular pollution and dumping of garbage by massive tourist influx.

Ooty Lake (Tamil Nadu)   Lake is dying due to destsruction of aquatic life by massive tourist invasion and dumping of garbage and sewage.

GLOBAL HOT SPOTS

1. Amazon (Latin America)
2. Hudson Bay (US and Canada)
3. Arctic Tundra (North Pole)
4. River Rhine (Switzerland)
5. Brittany Sea Shore (France)
6. Alaska (USA)
7. Island of Madagascar (East African Coast)
8. St.Helena Island
9. Lord Howe Island
10. Seychelles Island
11. Masai Mara (Kenya)
12. Alps (Europe)
13. Maldives Island (Southeast Asia)
14. Caribbean Island (South Pacific)
15. The Mediterranean Sea Coast
16. Mauritius (East African Coast)
17. Calauit Island (Philippines)
18. Antarctica (South Pole)
19. Lake Victoria (Kenya)

THREATS TO BIODIVERSITY

Species are now dying out as fast as it did since the mass extinction at the end of Cretaceous period, some 65 million years ago. Tropical deforestation is the cause of today's extinction crisis. About 17 million hectares of tropical forests are cleared every year. If this trend continues, at least 5–10 per cent of tropical forest species will face extinction in the next 30 years.

The list of species at risk is swelling. Some 633 species are listed as being in danger of extinction in the United States alone, and more than 3,000 other species have been proposed for listing. In the past several hundred years, 80 species are believed to have become extinct in the continental United States and Hawaii; and a further 210 species are likely to be extinct (Figure 12.1). One-third of North America's freshwater fish are rare, threatened or endangered. Worldwide, more than 700 extinctions of vertebrates, invertebrates and vascular plants have been recorded since 1600. In Indonesia, 1500 local rice varieties have become extinct in the past 15 years. Worldwide, some 492 genetically distinct populations of tree species are endangered. In north-western United States, 169 genetically distinct populations of anadromous fish are facing the risk of extinction.

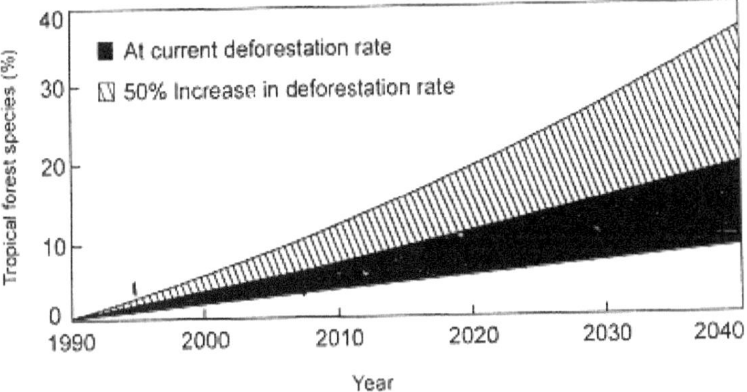

Figure 12.1   Percentage of tropical forest species likely to become extinct

Though biological extinction has been a natural phenomenon in geological history, the rate of extinction was perhaps one species per 1,000 years. However, between 1600 and 1950, the rate increased to one every ten years and currently it may be one every year. According to an estimate of Forest Survey of India, only 640,164 sq. km of land of forest is left now out of a 4,017,009 sq.km. Faunal losses have been mainly due to over exploitation, habitat alteration and pollution. The other possible reasons for the loss of species could be improper use of agro-chemicals and pesticides, a rapidly growing human population, inequitable land distribution and economic and political policies and constraints.

However, extinction is not caused by a single reason but by the cumulative effect of a host of causes. The Earth's biodiversity being one big interlink, the loss of one species in most cases triggers off the downward slide and consequent loss of many other dependent species.

Decline in genetic diversity in agriculture is costly as well. The Irish potato famine in 1846, the Soviet wheat crop loss in 1972, citrus canker in Brazil in 1991, all have been the result of loss of genetic diversity.

We pay high cost for the loss of a species. The water we drink, the air we breathe, our fertile soils, and our productive seas are all products of healthy biological systems. The loss of ecosystems, for instance, wetlands that provide critical services such as flood control, fish production, and pollutant assimilation is a direct economic cost as well as threat to species survival. Loss of biodiversity also destroys opportunities of recreation and tourism business—now worth as much as $12 billion a year worldwide.

CAUSES OF EXTINCTION

Population risk   Random variations in population rates (i.e., birth and death rates) cause a species in low abundance to become extinct.

Environmental risk   Environmental risk means variations in the physical or biological environment, including variations in predator, prey symbiotic or competitor species.

Natural catastrophe   Fires, storms, floods, earthquakes, volcanic eruptions, changes in oceanic currents and upwelling, etc. are the natural catastrophe.

Genetic risk   Determine change in genetic characteristic has small population of a species, due to reduced genetic variation or mutation, makes the species more vulnerable to extinction.

Human causes   Habitat loss and destruction inevitably results from the expansion of human populations and human activities. The ever-expanding human settlements have been causing destruction of natural ecosystems to meet their requirements of food, space, shelter, etc. The greatest destruction of biological communities has occurred during the last 150 years during which the human population went from just one billion in 1850 to 6.2 billion in 2002 and will reach an estimated 7.8 billion in 2025. In many countries, particularly Islands and where human population density is high, most of the original habitat has already been destroyed.

Destruction of forests, wetlands, grasslands, mangroves and other biological rich ecosystems around the world threatens to eliminate thousands or perhaps millions of species in a human-caused mass extinction that could rival those of geological history.

A few examples are listed below.

◇ The loss of local plant races for rice, wheat and maize has been a particularly serious problem in various parts of the world.

◇ The introduction of new high-yielding varieties of wheat has caused severe genetic erosion in Turkey, Iraq, Afganistan and Pakistan and India.

◇ 90% of the native varieties of wheat have been lost in 40 years in Greece.

◇ Some of the most valuable commercial species, particularly timber species in Ghana have suffered genetic erosion because of heavy exploitation and they are threatened with extinction in their areas of natural distribution.

◇ For the international market Rattan's (Lianoid palms) provide the second most important source of export earning from tropical forests of south and southest Asia. Rattans are mainly used in the production of cane furniture for international market whereas for local uses it is used for production of mats, baskets, fish traps dyes and medicines.

◇ Heavy exploitation combined with habitat destruction has led to the decline of major commercial Rattan species.

◇ Medicinal plants species to a large extent are harvested from the wild and relatively few are cultivated as crop plants.

◇ In international trade nearly five million number of plants belonging to 5,000 orchid species have been recorded in CITES.

◇ Thailand is a major source country for Orchids in international trade, and *Dendrobium* is the most heavily traded orchid genus exported.

◇ In Japan, out of 70 orchid taxa enlisted in the Japanese Red data book 50 are threatened by over collection.

◇   Orchid collection together with habitat destruction has led to the decline of wild orchid species in many other countries.

◇   Cacti and other succulent plants including species of *Aloe, Euphorbia, Pachypodium* are also traded internationally, Netherlands, Mexico, and Madagascar are the main source countries which have exported cacti and succulent plants. This trade poses a severe treat to such plant species.

Habitat fragmentation   It is a process where a large, continuous area of habitat is both, reduced in area and divided into two or more fragments. Habitat fragmentation may take place due to the development of roads, towers canals, fields, industries, etc. in an original large habitat. The fragments thus formed are oftenly isolated from one another by highly modified or degraded landscapes. These isolated, small, scattered populations are susceptible to environmental hardships, and consequently, in the end, possible extinction.

Disease   Pathogens, or disease organisms, may also be considered predators. The incidence of disease in wild species may increase due to human activities. The extent of diseases increases further when animals are kept in captivity (even in sanctuaries or reserves) rather than being able to disperse over a large area. Animals, as compared to plants, are more prone to infection when they are under stress.

Genetic assimilation   Some rare and endangered species are threatened by genetic assimilation because they cross-breed with closely related species that are more numerous or more vigorous.

Pollution   Environmental pollution is the most subtle form of habitat degradation. The most common causes of which are pesticides, industrial effluents and emissions, and emission from automobiles. Toxic pollutants can have disastrous effects on local populations of organisms.

Poaching   Poaching is another insidious threat that has emerged in recent decades as one of the primary reasons for the decline in number of species. Poaching pressures, however, are unevenly distributed since certain selected species are more heavily targeted that others are. Despite legal protection in many countries, products from endangered species are widely traded within and between nations.

Wildlife is sold for live specimens, folk medicines, furs, hides; skin (or leather) and other products such as ivory, antlers and horns amounting to millions of dollors each year. Developing countries in Asia, Africa and Latin America with the richest biodiversity in the world are the main source of wild animal and animal products, while Europe, North America and some wealthy Asian countries are the principal importers.

Introduction of exotic species   Organisms introduced into habitats where they are not native are termed as exotics. They can be thought of as biological pollutants and are considered to be among the most damaging agents of habitat alteration and degradation in the world.

Inducing species intentionally or unintentionally (accidentally) from one habitat into another where they have never been before is a very risky business. Freed from the parasites, pathogens, predators and competitors that normally keep their numbers in check, exotics often exhibit explosive population growth that crowds out native species.

WILDLIFE

Wildlife refers to all free, undomesticated species. Sometimes the term is used to describe only free, undomesticated animal species.

The marshy lowlands of Assam and north Bengal are the native of the one horned Rhinoceros. In the hilly and forested areas of Ghats are found the Indian Bison, wild buffalo and Nilgai. Deer and antelope are also found in these forests. Numerous monkeys are found in the less dense forests and jungle, and a variety of fish are found in our rivers, lakes and tanks. India also has a number of species of fish, turtles. Crocodiles are found in large perennial rivers. There are species of colourful birds also. Pheasant, ducks, parakeets, cranes are the other common birds.

Conservation of Wildlife

Forests are being cut down for cultivation, mining, industry: for the coming up of cities and to make way for new roads. There has been a destruction of wildlife habitats. Past climatic changes have also led to the extinction of some species of wild animals.

It is necessary to prevent the extinction of wild plants and animals to retain the diversity of species in our country. "Indian Board of wildlife" has taken steps to prevent the extinction of animals by:

1. Setting up of sanctuaries and parks. (At present we have 96 parks and 421 sanctuaries besides 21 tiger reserves (Table 12.4)).

2. Appointment of observers for wild life.

3. Creation of buffer belts around the sanctuaries where animals are prohibited

Table 12.4 Sanctuaries of India

State	Site	Animals, birds
Assam	Kaziranga	Rhinoceros.
Bihar	Hazaribag	Tiger, leopard,
	National park	Nilgai, Sambar
	Palamau	Gaur
	National Park	
Gujarat	Gir sanctuary	Indian lion
Karnataka	Bandipur	Elephant, Gaur,
	Ranganthittu	Birds
	Bird Sanctuary	
Kashmir	Dachigam	Kashmir stag
Kerala	Periyar game	Elephant, gaur
		sanctuary
Madhya Pradesh	Kanha National Park	Tiger, Barasinga,
		Black buck
Maharashtra	Taroba National Park	Tiger, Bear
Rajasthan	Bharatpur sanctuary	Birds
Uttar Pradesh	Corbett National Park	Tiger, Elephant
Tamil Nadu	Mudumalai Sanctuary	Elephant
	Vedanthangal	Birds

Man–Animal Conflicts

Man is afraid of animals and animals are afraid of man. There is conflict between humans and elephants once found in all parts of India, except those with low rainfall, elephants are now restricted to the hill forests of Kerala, Tamil Nadu and Karnataka, the Northeast, the Himalayan foothills in UP and in West Bengal, Orissa and south Bihar. They have disappeared from Maharashtra, Madhya Pradesh and Andhra Pradesh which still possess suitable forest habitats.

In the Periyar Forest reserve in Kerala, once India's finest elephant habitat, tuskers that were easily seen in the fifties are now rare.

While the hunting-gathering tribes influenced life around them only mildly, the impact of human activity increased dramatically when people began to cultivate plants and maintain herds of domestic animals, beginning with the river valleys of north-western India some 5000 to 8000 years ago.

The cultivators' conflict extended beyond wild animals that attacked people, livestock, and crops to wild plants that competed with cultivated crops. As agriculturalists and pastoralists, our ancestors profoundly modified the habitat around them.

More insidious than hunting, however, was the destruction of forests wrought by the British because of the great demand for railway sleepers, wood charcoal, ship and so on. To get this wood, the British took over, without compensation, vast tracts of communally-owned forests throughout the country.

The range of impacts that humans have had on animals can be grouped conveniently into five main categories:

1.   domestication,
2.   dispersal,
3.   contraction,
4.   expansion and
5.   extinction.

Domestication of animals Humans have affected animals. In particular, humans have changed and enhanced the characteristics for which they originally chose to domesticate animals. For example, the wild ancestors of cattle gave no more than few hundred millilitres of milk; today the best milk cow can yield up to 15,000 litres of milk during its lactation period.

Indeed, one of the most important consequences or manifestations of the domestication of animals consists of a sharp change in the seasonal biology.

Dispersal and invasions of animals   There have been many accidental introductions, especially since the development of ocean-going vessels.

Some animals arrive accidentally with other beasts that are imported deliberately. There are great many examples of cattle, horses, donkeys and goats which have effectively adapted to new environments and have virtually become wild. Frequently they have both ousted native animals; and they are also a cause of desertification, as encountered in ocean inlands.

Animal contractions and decline   The extreme effect of human interference with animals is extinction, but before that point is reached humans may cause major contractions in both animal numbers and animal distribution. This decline may be brought about partly by intentional killings for subsistence and commercial purposes, but much wildlife decline occurs indirectly, for example, through pollution.

Heavy metals and methyl mercury may be built up in marine organisms, and filter feeders like shellfish have a strong tendency to concentrate the metals from very dilute solutions.

Oil pollution is an increasingly serious problem for marine and coastal fauna and flora, although some of it derives from natural seeps.

Purposeful hunting, particularly when it involves modem firearms, means that the distribution of many animal species has become smaller very rapidly.

The introduction of a new animal species can cause the decline of another, whether by predation, competition or by hybridization.

Expansion of animal populations   Human actions, however, are not invariably detrimental, and even great cities may have effects on animal life which can be considered desirable or tolerable. Human economic activities may lead to a rise in the number of examples of a particular habitat which can lead to an expansion in the distribution of certain species. Very often human activities do not lead to species diversity, but to important increases in numbers of individuals by creating new and favourable environments.

Animal extinctions   As the size of human populations has increased and technology developed, humans have been responsible for the extinction of many species of both birds and mammals. Extinction takes place relatively slowly after a long history of human activity like hunting, climatic analysis of loss of habitat and competition for resources.

One of the most fundamental ways in which humans are causing extinction is reducing the area of natural habitat available to a species. Reduction in area leads to reduction in numbers and this, in turn, can lead to genetic impoverishment through in breeding.

ENDANGERED AND ENDEMIC SPECIES

Wild species with so few individual survivors that the species could soon become extinct in all or most of the natural ranges are called endangered species.

Species that are found in only one area and are especially vulnerable to extinction are called endemic species.

The lion-tailed macaque which inhabits the evergreen rainforests and "sholas" of south India, is among the world's most endangered primates (Figure 12.2).

Figure 12.2   The lion-tailed macaque in evergreen forests

Eleven species of flying squirrels and two marmots belonging to the order Rodent are also on the endangered list. They are killed for their pelts and flesh. Indian reptiles which include lizards, snakes, crocodilians, tortoises and turtles are exploited for their eggs, flesh, skin and even venom. Granite block and embankments built as a protection against erosion of beach fronts have "fenced-off" 200 km of the Kerala coast from nesting sea turtles. This has led to the disappearance of four out of the five species of turtles found on the Kerala coast. On the east coast the Olive Ridley sea turtles are being slaughtered for their flesh and eggs.

Habitat destruction, pollution of air and water, and the increasing drainage of wetlands have taken a heavy toll of bird life in India. Hunting for sport and trapping for the cage bird trade have contributed to their decline. Deforestation threatens among others most species of hornbills, the bamboo partridge and Himalayan forests birds including colourful pheasants and tragopan species. The Indian Skimmer, a resident of the major Indian rivers, is becoming exceedingly rare due to water pollution.

The famous Dal lake in Srinagar is shrinking due to human encroachments. The Pulicat lake, a great centre for breeding of water birds in Andhra Pradesh, is threatened with chemical pollution. According to the Sixth Five Year Plan (1980–85) document, "Coral reefs rich in limestone have been thoughtlessly plundered for the manufacture of cement in the Gulf of Mannar (Tamil Nadu) and the Pirotan Islands (Gujarat), and threatened in Lakshadweep and in the Andaman Islands.

Table 12.5   List of Endangered and endemic species

Mammals

Andaman Wild Pig	Himalayan Tahr	Musk Deer
Bharal	Hispid Hare	Nilgiri Tahr
Binturang	Hog Badger	Ovis Ammon or Nyan
Black Buck	Hoolock Gibbon	Pallas's Cat
Blue Whale	Hump-backed Whale	Pangolin
Brow-antlered Deer	Indian Elephant	Pygmy Hog
Capped Langur	Indian Lion	Ratel
Caracal	Indian Wild Ass	Rhinoceros
Cetacean species	Indian Wolf	Rusty-spotted Cat
Cheetah	Kashmir Stag	Serow
Chinese Pangolin	Leaf Monkey	Sloth Bear
Chinkara or Indian Gazelle	Leopard or Panther	Slow Loris

Mammals

Clouded Leopard	Leopard Cat	Small Travancore Flying Squirrel
Crab-eating Macaque Desert Cat	Lesser or Red Panda	Snow Leopard
Dugong	Lion-tailed Macaque	Spotted Linsang
Fishing Cat	Loris	Swamp Deer
Four-Horned antelope	Lynx	Takin or Mishmi Takin
Dolphin (Gangetic)	Malabar Civet	Tibetan Antelope or Chiru Tibetan Gazelle
Flying Squirrels	Malay or Sun Bear	Tibetan Wild Ass
Golden Cat	Marbled Cat	Tiger
Golden Langur	Markhor	Urial or Shapu
Himalayan Ibex	Mouse Deer	Wild Buffalo
	Himalayan Tahr	Wild Yak

Amphibians and reptiles

Agra Monitor Lizard	Hawksbill Turtle	Olive-back logger-head Turtle
Atlantic Ridley Turtle	Himalayan Newt or Salamander	Peacock-marked soft-shelled Turtle
Barred, Oval or Yellow Monitor Lizard	Indian egg-eating Snake	Pythons
Crocodiles	Indian soft-shelled Turtle	Three-keeled Turtle
Gharial	Indian Tent Turtle	Tortoise
Ganges soft-shelled Turtle	Large Bengal Monitor Lizard Leathery Turtle	Viviparous Toad
Green Sea Turtle	Logger-head Turtle	Water lizard

Birds

Andaman Teal	Hooded Crane	Nicobar Pigeon
Assam bamboo Partridge	Hornbills	Osprey or Fish eating Eagle
Bazas	Houbara Bustard	Peacock Pheasant
Bengal Florican	Humes bar-backed Pheasant	Peafowl
Black-necked Crane	Indian Pied Hornbill	Pink-headed Duck
Blood Pheasants	Jerdon's Courser	Scalater's Monal
Brown headed Gull	lammergeier	Siberian White Crane
Cheer Pheasant	large Falcons	Spur Fowl
Eastern White Stork	large Whistling Teal	Tibetan Snow-cock
Forest spotted Owlet	Monal Pheasants	Tragopan Pheasants
Great Indian Bustard	Mountain Quail	White-bellied Sea Eagle
Great Indian Hornbill	Narcondom Hornbill	White-eared Pheasant
Hawks	Nicobar Megapode	White spoonbill
		White-winged Wood Duck

In Kerala, coastal evergreen forests now survive only in small pockets dedicated as sacred groves to serpent deities.

PROJECT TIGER

Project Tiger was initiated in 1973, in response to the alarming decrease in the population of wild tigers from 40,000 at the turn of the century down to 1827 in 1972. The project was launched by the Government of India with a grant of Rs.50 million in cooperation with WWF-India and International Union for Conservation of Nature (IUCN) who together pledged one million dollars for equipment and experts. In 1973, nine tiger reserves in nine states with a total area of 13,017 sq km were set aside, with a tiger population of 268. Today there are 11 reserves spread over 15,760 sq km in 10 states.

To control the illegal trade of endangered species and skins, India signed the Convention on International Trade in Endangered Species of Wild Fauna and Flora (CITES) in 1976. Despite a few big seizures in the last couple of years, international smuggling rings continue to flourish. Poor people in the countryside are induced to kill animals whose skins are purchased at the nominal price dealers. Compulsory checking of such cargo, has little deterrent effect.

The blackbuck is a species indigenous to India. Its thorn scrub forest habitat in arid and semi-arid regions has suffered the maximum level of destruction (Sharad Gaur).

A major effort in this direction is the Biosphere Reserve Programme, launched on a worldwide scale by UNESCO in 1973, as the Man and Biosphere (MAB) Programme. It has led to the creation of over 200 biospheres in about 40 countries so far. The Sixth Plan has directed the Department of Environment (DOE) to constitute a cadre of specialists for management of the Biosphere Reserves.

Eight areas are being considered for setting up such reserves. They include:

1.   Nilgiri at the trijunction of Karnataka, Kerala and Tamil Nadu
2.   Tura citrus gene sanctuary in Meghalaya
3.   Namdapha in Arunachal Pradesh
4.   Valley of Flowers
5.   Tunganath-Rudranath in UP
6.   Nanda Devi in UP
7.   Some islands in the Andaman-Nicobar group and
8.   Gulf of Mannar on the Tamil Nadu coast.

BIODIVERSITY CONSERVATION

Conservation is defined as "the management of human use of the biosphere so that it may yield the greatest sustainable benefit to the present generation while maintaining its potential to meet the needs and aspirations of future generations" (Figure 12.3). Conservation of our natural resources has the following three specific objectives:

1.   to maintain essential ecological processes and life-supporting systems;

2.   to preserve the diversity of species or the range of genetic material found in the organisms on the planet; and

3.   to ensure sustainable utilization of species and ecosystems which support millions of rural communities as well as the major industries all over the world.

The wildlife conservation efforts are mostly centred on protecting plant and animal life in protected habitats, such as botanical gardens, zoos, sanctuaries, national parks, biosphere reserves, etc. The two basic approaches to the wildlife conservation in protected habitats are:

1.   *in-situ* conservation and

2.   *ex-situ* conservation

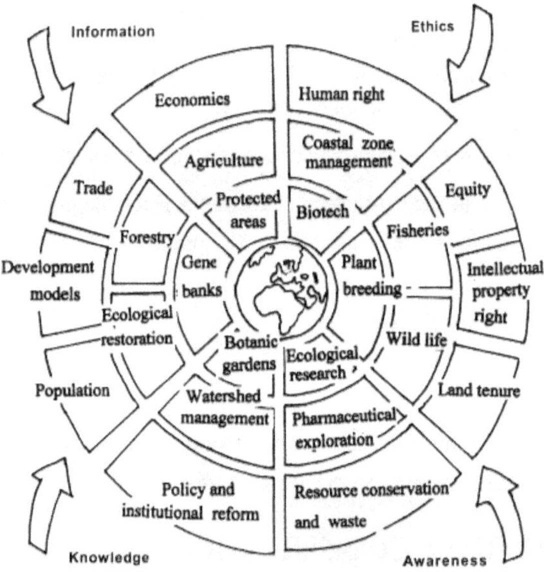

Figure 12.3   The scope of biodiversity conservation

Measures Taken in India for Conservation of Biodiversity

India is one of the major centres of biodiversity, because of the immense variety in physiography and the range of climatic conditions, which results in a diversity of ecological habitats ranging from tropical, sub-tropical, temperate and alpine to desert. As a result, according to the world biogeographic classification, India represents two of the major realms and three basic biomes, including thirteen biogeographic regions. The biosphere reserves are: the Nilgiris (Tamil Nadu, Kerala and Karnataka), Namdapha (Arunachal Pradesh), Nanda Devi (Uttar Pradesh), Valley of Flowers (Uttar Pradesh), Great Nicobar, Gulf of Mannar (Tamil Nadu), Kaziranga (Assam), Manas (Assam), Sunderbans (West Bengal), Thar Desert (Rajasthan), Kanha (Madhya Pradesh), Nokrek Tura (Meghalaya) and the Rann of Kutch (Gujarat). The Gulf of Mannar is a marine biosphere reserve. Recently, this classification has been modified by the Wildlife Institute of India, which divides the country into ten biogeographic regions:

1. Trans-Himalayan
2. Himalayan
3. Indian Desert
4. Semi-Arid
5. Western Ghats
6. Deccan Peninsula
7. Gangetic Plain
8. Northeast India
9. Islands and
10. Coasts

The Government of India has numerous and wide-ranging policies, programmes and projects, which directly or indirectly serve to protect and conserve the country's biological diversity.

Legal Framework

The constitution of India ensures the concept of environmental protection. It has a new section on Directive Principles of State Policy, setting out the duties for the states and all the citizens through Article 48A and Article 51A (g) and explaining that "the State shall endeavour to protect and improve the environment and "to safeguard the forests and wildlife in the country" and to protect and improve the natural environment, including forests, lakes and rivers and wildlife, and to have compassion for living creatures."

A legal and policy framework has been developed which relates specifically to biological diversity. The Forest Act of 1927, the Forest (Conservation) Act of 1980 (amended in 1988), the Wildlife (Protection) Act of 1972 (amended in1983, 1986 and 1991) and the Environment (Protection) Act of 1986 are all supported by state laws and statutes relating to forests and other natural resources.

The National Conservation Strategy of 1992 outlines the policy actions required, giving greater attention to biodiversity conservation. The National Forest Policy, as amended in 1988, stresses the sustainable use of forests, and the need for greater attention to ecologically fragile (but biologically rich) areas such as the mountain and island ecosystems. The National Wildlife Action Plan of 1973 lays down priorities in the area of wildlife conservation. One of the major considerations in Environmental Impact Assessment carried out by the Ministry of Environment and Forests is the protection of habitats and valuable ecosystems.

Surveys

Surveys of the floral and faunal resources in India are carried out by the Botanical Survey of India (BSI). The Forest Survey of India (FSD) uses satellite imagery, aerial photography and ground verification to assess forest and tree cover with a view to developing an accurate database for planning and monitoring purposes.

In situ Conservation

Conservation of a species in its natural ecosystem or in a man-made ecosystem is called *in situ* conservation.

India has a long history of *insitu* conservation of fauna through protected areas. Protected areas include national parks, sanctuaries and biosphere reserves. With the setting up of the Indian Board for Wildlife in 1952 and the enactment of the Wildlife (Protection) Act of 1972, the protected areas network has been strengthened. Today, there are 73 national parks and 413 sanctuaries covering about 4% of the total geographic area of the country. Project Tiger, launched in 1973, succeeded in increasing the tiger population to more than 4,000 in 1989, spread over eighteen Tiger Reserves in a total area of 28,000 km². The Wildlife Institute of India has recommended that the number of protected areas should be increased to 148 national parks and 503 sanctuaries, covering 4.6% of the total area of the country. However, the critical problem is not merely the conservation of particular species or habitats. It is to ensure the unhindered evolution of microorganisms, plants and animals in their totality, and as part of the natural ecosystems.

In recent years, programmes have also been launched for ecological research and for the protection of fragile ecosystems, including mangroves, wetlands and coral reefs. A National Committee on wetlands, mangroves and coral reefs, established by the Ministry of Environment and Forests, has identified fifteen mangrove areas, sixteen wetlands and four coral reef areas (Gulf of Mannar, Kutch, the Andaman and Nicobar and Lakshadweep Islands) for integrated management and their conservation.

Ex situ Conservation

Conservation of species particularly of endangered species, away from natural habitat under human supervision is called *ex situ* conservation.

In order to complement the efforts made for *in situ* conservation, attention has also been given to *ex situ* conservation. There are 33 botanical gardens and 33 university-level biological gardens. There are also 205 areas for *ex situ* wildlife preservation, including 107 zoos, 49 deer parks, thirteen safari parks, six snake parks, 24 nature/education/breeding centres and six aquaria. Some of the major zoos have made significant achievements in the captive breeding of endangered species. To improve the management of zoos, the Government of India has recently set up the Central Zoo Authority.

The collection and preservation of genetic resources are done through the National Bureau of Plant Genetic Resources in Delhi (for the wild relatives of crop plants), the National Bureau of Animal Genetic Resources in Karnal (for domestic animals) and the National Bureau of Fish Genetic Resources in Allahabad (for economically valuable fish). These bureaux are assigned the task of collecting germplasm from within and outside India, and also of supplying it on request to Indian and foreign agencies for research purposes.

The Department of Biotechnology was established in 1986 to guide, supervise and develop biotechnology programmes in the country and to develop the infrastructural support for this purpose.

Some important facilities developed under this programme are National facilities for the animal houses, genetic engineering units, biochemical engineering units, immunology and biosafety laboratories, and culture of microbes, blue-green algae, marine cyanobacteria and plant tissues.

Recommendations

India has wealth in the form of rich biodiversity, which has been fulfilling the needs of the people. We cannot afford any more exploitation of existing biological resources in an unsustainable manner. Now the time is ripe for preparing our people for *in situ* and *ex situ* conservation of the existing biodiversity. Both are essential and complementary to each other. The author recommends that:

i.   Environmental education should be imparted to the people and to those who are directly dependent upon biomass.

ii. Merely declaring an area as a sanctuary, Project Tiger or National Park will not suffice. We should believe in people's participation. We have to free the people from their dependence on biomass.

iii. The hands of the NGOs should be further strengthened so that with more financial assistance they can help educate people better about the benefits of biodiversity and the need for *in situ* and *ex situ* conservation of biodiversity.

iv. The practice of removing indigenous species and replacing them with monocultures of exotic species should be discouraged.

v.   Environmental ethics and religious means should be used to conserve species and genetic diversity.

vi. People should be made aware of the importance of their local environmental and to degradation.

vii. Environmental Impact Assessment should be mandatory to check the erosion of ecological refuges, deforestation and loss of genetic diversity.

Levels of Action

Biodiversity conservation is gaining ground precisely because the issue is so vast that it can encompass the interest of so many countries. Four major steps have been taken to shape an international response to the loss of biodiversity, to support the actions already underway at local, regional and national levels.

i. *Global Environment Facility (GEF)*   World Bank, United Nations Development Programme, United Nations Environment Programme established the Global Environment Facility in 1991 on a three-year pilot basis. The GEF is expected to commit $400 million to conserve biodiversity.

ii. *International Biodiversity Strategy Programme*   World Resources Institute, World Conservation Union, UNEP and more than 40 governmental and non-governmental organizations outlined the programmes to stop the loss of biodiversity and mobilize its benefits to human needs sustainably and equitably.

iii. *Convention on Biological Diversity*   Under the auspices of UNEP, more than 100 nations met during the Earth Summit at Brazil to establish a legal framework governing international financial support for biodiversity conservation, the identification of international conservation priorities and technology transfer for conservation and use of biodiversity.

iv. *Agenda* 21   Developed through a series of inter-governmental preparatory meetings with input from a variety of non-governmental processes including the Biodiversity Strategy Programme-AGENDA 21 provides a plan of action on a number of issues including biodiversity.

Promising moves are underway to slow down the loss of biodiversity and mobilize its benefits for human well-being. Many of today's positive initiatives may turn out to be stop-gap arrangements unless advances are made in dealing with over consumption of resources, population growth, misguided resource management policies and social and economic inequities. The need to conserve our planet's biological wealth gives nations even more impetus to solve these knotty problems. At the same time, sustaining biodiversity may help provide the means for crafting solutions.

REVIEW QUESTIONS

1. Explain the term ecosystem. How is the ecosystem classified.
2. Explain the structure and function of the ecosystem.
3. Explain the flow of energy in an ecosystem.
4. What is ecological succession? Name the difference types of ecological succession and describe.
5. What are food chain and food web? Describe their significance.
6. What are Eltonian pyramids? Describe the different types. What are their significance?
7. Describe the characteristic features of various types of ecosystems. Describe the structure and function of forest ecosystem.
8. Write notes on
   i. Pond ecosystem
   ii. Desert ecosystem
   iii. Grassland ecosystem
   iv. Ocean ecosystem
9. What is lentic and lotic ecosystem? Describe the abiotic and biotic components on any fresh water ecosystem.
10. Describe the characteristic feature of grassland ecosystem.
11. What do you understand by biodiversity? What are the different levels of biodiversity? At these levels, explain biodiversity?
12. What is the bio-geo-classification of India? Explain its importance.
13. What are direct and indirect values of biodiversity.
14. Explain the uses of biodiversity.
15. Comment on India as a major diversity nation.

16. What are the hot spots of biodiversity? Name a few hot spots of biodiversity at the global level.

17. Name two hot spots of biodiversity in India and explain any one.

18. Explain global patterns in biodiversity.

19. Discuss the causes of extinction of species.

20. How does extinction of species affect biodiversity. What are the causes of extinction?

21. Write notes on:

    i.   Endangered species of India

    ii. Endemic species of India

22. What are the objectives of conservation of biodiversity? Name any one approach to wildlife conservation and explain.

23. Explain *in situ* conservation of biodiversity and discuss the merits and demerits.

24. What is meant by *ex situ* conservation and discuss the merits and demerits.

INTRODUCTION

Environmental pollution is a term that refers to all the ways that human activity harms the natural environment. Most people have witnessed environmental pollution in the form of an open garbage dump or an automobile pouring out black smoke. However, pollution can also be invisible, odourless, and tasteless. Some kinds of pollution do not actually dirty the land, air, or water, but they reduce the quality of life of people and other living things. For example, noise from traffic and machinery can be considered as forms of pollution.

Environmental pollution is one of the most serious problems facing humanity and other life forms today. Badly polluted air can harm crops and cause life-threatening illnesses. Some air pollutants have reduced the capacity of the atmosphere to filter out the sun's harmful ultraviolet radiation. Most scientists believe that these and other air pollutants have begun to change climates around the world. Water and soil pollution threaten the ability of farmers to grow enough food. Ocean pollution endangers many marine organisms.

Air pollution has increased alarmingly due to population explosion, industrialization, urbanization, automobiles and other human pro-activities for greater comfort. The atmosphere is a dynamic system, which steadily absorbs various pollutants from natural as well as man-made sources, thus acting as a natural sink. Gases such as CO, CO_2, H_2S, SO_2 and NO_x as well as particulate matter such as sand and dust are continuously released into the atmosphere through natural activities such as forest fires, decay of vegetation, winds, and sand or dust forms and man-made activities. The pollutants contributed by man-made activities have surpassed that contributed by nature a thousand-fold.

These pollutants travel through the air, disperse and may interact with other substances in the atmosphere before they reach a sink such as an ocean or a human receptor. If the pollutants enter the atmosphere at a faster rate than that are absorbed by the natural sinks, then they gradually accumulate in the air. Such a disturbance in the dynamic equilibrium in the atmosphere by the air pollutants released by anthropogenic activities resulting in considerable accumulation in the atmosphere may effect the very life on earth and its environment. Further, the dilution and dispersion of the gaseous pollutant in the atmosphere depend upon the meteorological conditions prevailing at a given time.

MAJOR CLASSES OF POLLUTANTS

The major classes of pollutants commonly found in the atmosphere are listed below in Table 13.1.

Table 13.1   Common pollutants in the atmosphere

Carbon oxides	Carbon monoxides (CO) and carbon dioxide (CO_2)
Sulphur oxides	Sulphur dioxide (SO_2) and sulphur trioxide (SO_3)
Nitrogen oxides	Nitric oxide (NO), nitrogen dioxide (NO_2), nitrous oxide (N_2O) (NO and NO_2 often are lumped together and labelled NO_x)
Volatile organic compounds (VOCs)	Methane (CH_4), propane (C_3H_8), chlorofluorocarbons (CFCs)
Suspended particulate matter (SPM)	Solid particles (dust, soot, asbestos, lead, nitrate, and sulphate salts), liquid droplets (sulphuric acid, PCBs, dioxins and pesticides)
Photochemical oxidants	Ozone (O_3), peroxy acyl nitrates (PANs), hydrogen peroxide (H_2O_2), aldehydes
Radioactive substances	Radon-222, iodine-131, strontium-90, plutonium-239
Hazardous air pollutants (HAPs), which cause health effects such as cancer, birth defects, and nervous system problems	Carbon tetrachloride (CCl_4), methyl chloride (CH_3Cl), chloroform ($CHCl_3$), benzene (C_6H_6), ethylene dibromide ($C_2H_2Br_2$), formaldehyde (CH_2O_2)

Most outdoor pollutants in urban areas enter the atmosphere from the burning of fossil fuels in power plants and factories (stationary sources) and in motor vehicles (mobile sources). Because such pollutants are concentrated in a localized area, they can build up to harmful levels.

Classification of Air Pollutants

Air pollutants are classified into two types, namely, primary pollutants and secondary pollutants (Figure 13.1).

Primary pollutants   The pollutants that are emitted directly into the air in a potentially harmful manner are called primary pollutants, e.g. carbon monoxide, nitric oxide, sulphur dioxide and hydrocarbons.

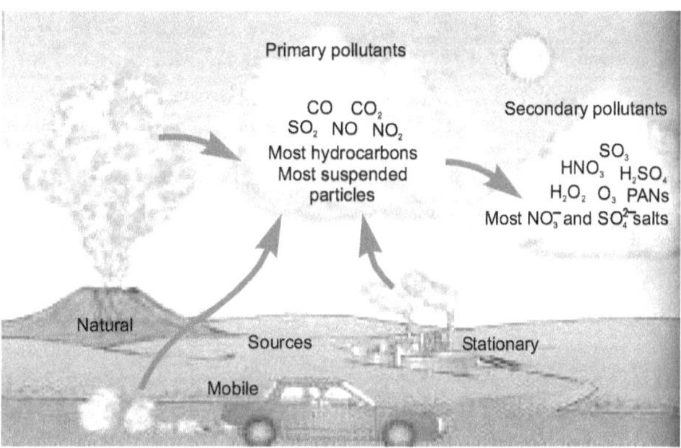

Figure 13.1   Pollutants

Secondary pollutants   The pollutants that are derived from the primary pollutants due to some chemical or photochemical reactions in the atmosphere are called the secondary pollutants, e.g. ozone, peroxy acyl nitrate (PAN), and photochemical smog.

Mostly urban areas have higher pollution levels than the rural areas because of increasing factories and cars. However, prevailing winds can spread the long-lived primary and secondary air pollutants that are strongly emitted in urban and industrial areas to the countryside and to the downward urban and rural areas.

According to the WHO, more than 1.1 billion people live in urban areas where ambient air is unhealthy to breathe.

The pollutants are further classified as gaseous pollutants and particulate pollutants based on the state of matter.

Gaseous pollutants   The pollutants that are mixed with the air and do not normally settle out are called gaseous pollutants, e.g. CO, NO_x and SO_2, etc.

Particulate pollutants   These are the pollutants that comprise of finely divided solids or liquids and often exist in colloidal state as aerosols, e.g. smoke, fumes, dust, fog, smog and sprays.

GASEOUS POLLUTANTS

Oxides of Carbon

Carbon monoxide (CO)   It is a colourless and odourless gas formed during the incomplete combustion of carbon-containing fuels. It is poisonous to breathing animals.

$$(2C + O_2 \rightarrow 2CO)$$

About 77%, comes from motor vehicle exhausts, whereas in cities it ranges to 95%. The major human sources are cigarette smoking and incomplete burning of fossil fuels.

Health effects   CO reacts with the haemoglobin in red blood cells and reduces the ability of blood to bring oxygen to body cells and tissues. This impairs perception and thinking, slows down the reflexes, causes headache, drowsiness, dizziness, and nausea, triggers heart attacks and angina, damages the development of foetuses, and aggravates chronic, bronchitis, emphyema, and anaemia. At high levels, it causes collapse, coma, irreversible brain cell damage and death.

Effects on plants   CO has some detrimental effects on plants, when exposed for long duration. For example, it inhibits the nitrogen fixation ability of bacteria when exposed for long duration say 33–38 hours. For example, the nitrogen-fixing ability of bacteria living in clover roots is inhibited to a 100-ppm concentration for 35 days. Since the CO levels hardly reach 100 ppm, it creates no significant effects on plants and vegetation. However, co-concentration from 100–10,000 ppm affects leaf drop, leaf curling, reduction in leaf size and chlorophyll with premature ageing.

In Bihar and also in Kazakhstan and neighbouring Central Asian States, people mix tobacco and lime known as khaini and place it under the tongue or in the groove between the teeth and the lip. Such people are prone to khaini cancer in those parts of the mouth where they keep the mixer of tobacco and lime for a long time.

Oxides of Nitrogen

The oxides of nitrogen involved in air pollution (NO_x) are N_2O, NO, NO_2, N_2O_3 and N_2O_5, of which only N_2O, NO and NO_2 are 0.25 ppm, 0.1 to 2 ppm and 0.5 to 4 ppm respectively. They are usually represented together as NO_x. About 95% of Nitrogen oxide is emitted as NO and the remaining 5% as NO_2. They are largely emitted by automobiles and electric power industries in developed countries. In metropolitan cities, vehicular exhaust is the most important source of nitrogen oxides. NO_2 is the strong absorber of UV light and the chief constituent of photochemical smog (Figure 13.2). It initiates photochemical reactions in troposphere. It is the main pollutant of the Los Angel's smog. When fuels are burnt in air, some of the N_2 in the air is oxidized to NO. The amount of NO formed depends on the temperature of this flame and on the rate of cooling on quenching of the combination products. In other words, high flame temperatures and rapid cooling of the combination products are responsible for the formation of NO.

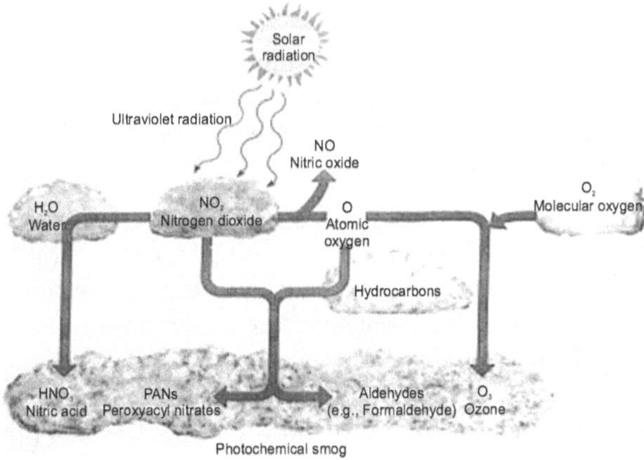

Figure 13.2   Simplified scheme of the formation of photochemical smog

The average residence times of NO and NO_2 in the atmosphere are 4 days and 3 days respectively. They undergo various photochemical and chemical reactions in the atmosphere, leading to the formation of HNO_3, which gets precipitated as nitrates during rainfall or as dust.

Sunlight-induced chemical and photochemical reactions involving NO_2 and hydrocarbon are responsible for the formation of photochemical smog. The NO_x from manmade sources may be 10 to 100 times more in urban areas when compared to rural areas. Even in the urban areas, the ambient NO_x levels vary with sunlight and traffic density at any given point of time.

Major human sources   It is obtained by fossil fuel burning in motor vehicles and power and industrial plants.

Health effects   It causes lung irritation and damage, aggravates asthma and chronic bronchitis, and increases the susceptibility to respiratory infections such as the flu and common colds (especially in young children and in older adults).

Environmental effects   It reduces the visibility. The acid deposition of HNO_3 can damage trees, soils and aquatic life in lakes.

Property damage HNO_3 can corrode the metals and eat away the stone on buildings, statues and monuments. NO_2 can damage fabrics.

Sulphur Dioxide (SO_2)

It is a colourless, irritating gas formed from the combustion of sulphur-containing fossil fuels such as coals and oil ($S+O_2 \rightarrow SO_2$). In the atmosphere, it can be converted to sulphuric acid (H_2SO_4)—a major component of acid deposition.

Source   Combustion of any sulphur-bearing materials produces SO_2 accompanied by a small quantity of SO_3. This mixture is normally denoted as SO_x. Nearly 67% of global SO_x pollution is due to volcanic activity and other natural sources, over which we have no control. The remaining 33% SO_x emission is due to human activities such as combustion of fuels, coal-fixed power stations, transportations, refineries, metallurgical operations such as smelting of sulphide ore and chemical plants, e.g. manufacture of sulphuric acid. Most of the man-made SO_x pollution is concentrated in urban and industrial areas.

Almost all the sulphur present in liquid and gaseous fuels and about 80% of sulphur present in the solid fuels appears as SO_x in the fuel gases. Depending on the sulphur content of the fuel burnt and on the conditions of combustion, the concentration of SO_x in fuel gases varies from 0.05–0.4%.

Smog SO_2 is oxidized to SO_3 in atmospheric air by photolytic and catalytic process involving ozone, NO_x and hydrocarbons, giving rise to the formation of photochemical smog. Oxidation of SO_2 can take place in the presence of catalysts such as NO_x, metal oxides, and ozone.

SO_2 combines with vapour to produce droplets of H_2SO_4 aerosol which gives rise to the so called acid rain. The sulphuric acid and sulphate aerosols present in the urban air are smaller than 2μ and hence can easily reach the pulmonary region of lungs, causing serious respiratory problems, particularly in old people.

$$SO_2 + O_3 \longrightarrow SO_3 + O_2$$

$$SO_3 + H_2O \longrightarrow H_2SO_4 \longrightarrow (H_2SO_4)_n$$

$$SO_2 + \tfrac{1}{2}O_2 + H_2O \longrightarrow H_2SO_4$$

Major human sources   It is obtained by burning coal in power plants (88%) and industrial processes (10%).

Health effects   Breathing problems for healthy people; restriction of airways in people with asthma; chronic exposure can cause a permanent condition similar to bronchitis. According to the WHO, at least 625 million people are exposed to unsafe levels of sulphur dioxide from fossil fuel burning.

Environmental effects   Reduces visibility; acid deposition of H_2SO_4 can damage trees, soils, and aquatic life in lakes.

Property damage SO_2 and H_2SO_4 can corrode metals and eat away stone on buildings, statues and monuments. SO_2 can damage paint, paper and leather.

Acid rain   Rain has always been highly valued by mankind, because good crops and abundant water supplies are possible only by timely and plentiful rainfall. Autumn rains and winter snow help cleaning the air. Over the last few decades, simple rainfall has taken on a threatening complexity in some parts of the world. In these localities, the rain must pass through an atmosphere polluted with oxides of sulphur and oxides of nitrogen. The falling rain and snow react with these oxide pollutants to produce often a mixture of sulphuric acid, nitric acid and water. This is known as acid precipitation or acid rain (Figure 13.3).

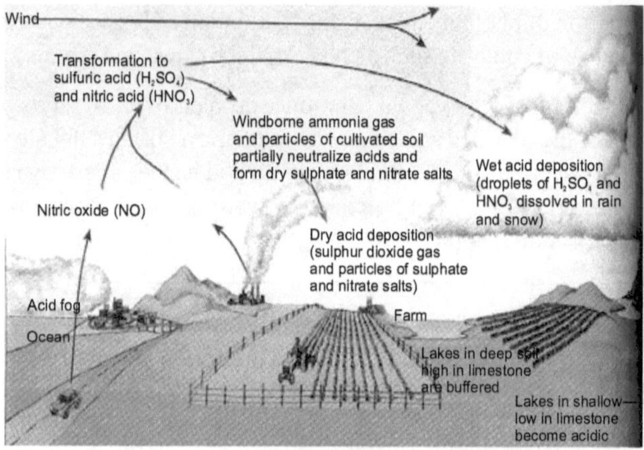

Figure 13.3   Acid rain

Normally unpolluted rain is weakly acidic and has a pH of 5.6 because CO_2 from the air reacts with water to form H_2CO_3. But acid rain that has a pH as low as 4 is ten times as acid as normal rain with a pH 5. Pure rainwater has a pH of about 5.5 to 5.7, but due to SO_2 emissions, the pH of rain can drop as low as 2.0. This increases the acidity of waterways particularly rivers and lakes.

Effects of acid rains   About 237 lakes in the Adirondack have a pH below 5, creating highly acidic level that is lethal for fish. There are 15,000 fishless lakes in Sweden and about 100 such lakes in the Adirondack region of USA, because of increased acidity of the lake and several rivers as Twyi are acidic. Many bacteria and blue green algae are killed due to the acidification, disrupting the whole ecological balance. Acid-rain killed fishes in lakes and destroyed trees in a wide swathe across Europe. The acid rain had damaged the leaves of plants and trees and had retarded the growth of Swedish forests. Effect of acid precipitation on terrestrial vegetation indicates a reduced rate of photosynthesis and growth and increased sensitivity to drought and disease. Acid lakes have low levels of phytoplankton. Snails, clams, oysters, etc. having their shells of calcium carbonate are among the first animals to perish in acidic lakes.

Acid rain causes extensive damage to buildings and structural materials of marble, limestone, slate and mortar, etc. Limestone is attached rapidly and this is termed as stone leprosy.

Presently the Taj Mahal in Agra is suffering due to SO_2 and H_2SO_4 acid fumes or air pollutants released from Mathura refinery. British Parliament building is also suffering damage due to H_2SO_4 rains. Ornamental stone work on a church in Bristol is corroded by acid in the air.

Acid rain is dangerous to human nervous system and respiratory system and thus humans are easy preys to neurological diseases.

Prevention and Clean-up of Acid Deposition

- Reduce the air pollution by improving the energy efficiency
- Reduce the coal use
- Increase the natural gas use
- Increase the use of renewable resource
- Burn the low-sulphur coal
- Remove the SO_2 particulates and NO_x from smokestack gases
- Remove the NO_x from motor vehicular exhaust
- Tax the emissions of SO_2
- Add lime to neutralize the acidified lakes
- Add phosphate fertilizer to neutralize the acidified lakes

SUSPENDED PARTICULATE MATTER (SPM)

Particulate pollutants are finally divided solid or liquid particles. Particles can be compared of inert or extremely reactive materials of size from 0.0002 (small molecule) to 500 microns (1 micron = 10^{-6} m), e.g. dust, smoke fumes. Based on the size, the particulate matter can be classified as follows.

Aerosols

Aerosols are nothing but air suspension. The particles suspended may be dust, smoke mist and fumes. This is a most general term applied to any tiny particles, either liquids or solids, dispersed in the atmosphere.

Dust

It is an air suspension of irregular-shaped mineral or other particles of size 1 to 200 microns. They settle under the influence of gravity. It is generated by crushing, chipping, grinding and by natural disintegration of rock and soil. Particles of this size range will be in the suspended state for a few seconds to several months. Particles of size less than 100 microns are called as fine dust whereas greater than 100 microns are called as coarse dust. Particles of size greater than 50 microns can be seen with unaided eye. The typical sizes of dust are given below.

Dust particles of crushing & Grinding	>20 microns
Fly ash from chimney	3–80 microns
Cement dust	10–150 microns
Foundry dust	1–200 microns

Smoke

It is an aerosol of fine carbon particles of size ranging from 0.5 to 1.0 microns, which are produced by incomplete combustion of organic particles such as coal, wood, etc.

Soot

Soot is the agglomeration of carbon particles of size 1 to 10 microns impregnated with tar formed due to incomplete combustion of carbon aerosols materials.

Fumes

Fumes are the fine solid particles formed by the condensation of gaseous state after vitalization. The size of particles ranges from 0.03 to 1 micron.

Mist

Mist is an aerosol of liquid droplets formed by the condensation of vapour. The size ranges of natural water vapor mists are from 40 to 500 microns. Usually the size is less than 10 μ

Fog

The mist formed out of droplets of water is called fog.

Smog

Smoke plus fog is expressed as smog

Spray

It consists of liquid droplets formed by the atomization of parent liquid, e.g. pesticides. The size of particles ranges from 10 to 1000 microns.

Haze

Haze is an air pollution condition formed due to the presence of very fine dust, mist, etc. in the atmosphere.

Major human sources   Burning coal in power and industrial plants (40%) burning diesel and other fuels in vehicles (17%), agriculture (plowing burning off fields), unpaved roads, and construction.

Health effects   It causes nose and throat irritation, lung damage and bronchitis; aggravates bronchitis and asthmas, shortens life, toxic particulates (such as lead cadmium, PCBs, and dioxins); and can cause mutations, reproductive problems, and cancer.

Property damage   Corrodes the metals and soil and discolours buildings, clothes, fabrics and paints.

Temperature inversions   During daylight, the sun warms the air near the earth's surface. Normally, this warm air having most of the pollutants rises to mix with the cooler air above it. This mixing of warm and cool air creates turbulence, which disperses the pollutants.

Under certain atmospheric conditions, however, a layer of warm air can lie atop a layer of cooler air nearer the ground, a situation known as a temperature inversion (Figure 13.4a and 13.4b). Because the cooler air is denser than the warmer air above it, the air near the surface does not rise and mix with the air above it. Pollutants can concentrate in this stagnant layer of cool air near the ground.

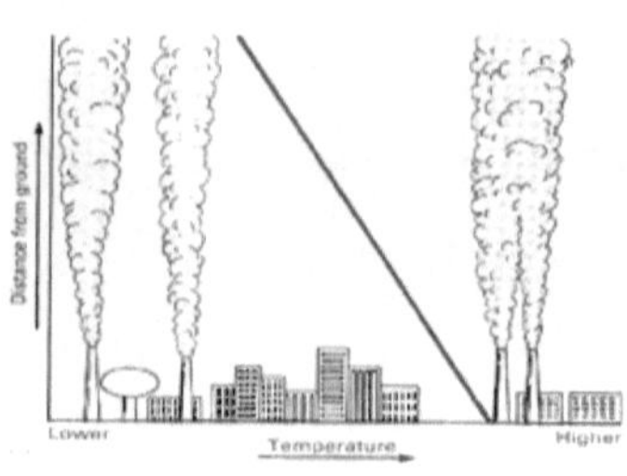

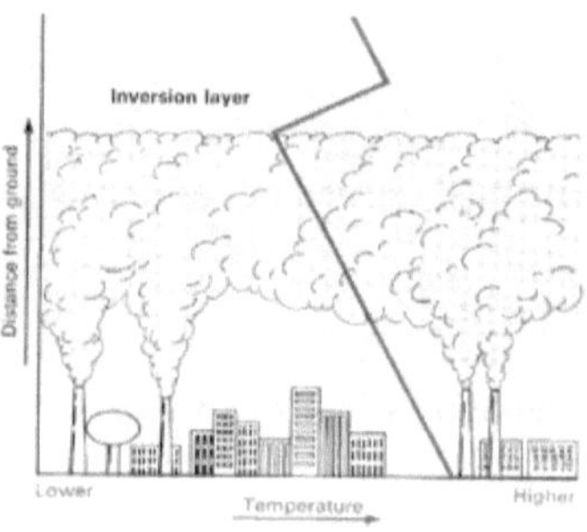

Figure 13.4a   Air temperature with altitude

Figure 13.4b   Thermal inversion layer

As long as these stagnant conditions persist, concentrations of pollutants in the valley below will build up to harmful and even lethal concentrations. This is what happened during the 1948, air pollution disaster in the valley town of Donora, Pennsylvania.

Case Study 16

Deterioration of Taj Mahal

Mathura Oil Refinery lies only 40 km away from Taj Mahal—one of the eight wonders of the world. This oil refinery emits about 25–30 tons of SO_2 daily inspite of using low-sulphur fuels. Air pollution surveys by Indian and International firms have estimated that any increase in SO_2 concentrations above the present 1.75 kg/m^3 would result in an acidic precipitation under conditions of low-wind speeds combined with humidity. This would result in the conversion of SO_2 into sulphuric acid—which is the main cause of "Stone cancer".

This sulphuric acid could react with calcium carbonate ($CaCO_3$) in the marble to form calcium sulphate ($CaSO_4$)—which would cause pitting in the Taj.

Discolouration of the white marble surface, i.e., appearance of a yellowish layer or yellow-grey deposits or brown rust-like stains on the white marble, especially in the niches and arches, chipping and breaking of the edges of the marble slabs, and formation of cracks in marble are some of the signs of deterioration of the Taj. Many steps are being taken to save the monument from the deteriorating effects of environmental pollution. They are as follows:

i.   Establishment of pollution-monitoring stations inside the Taj Mahal.

ii. Establishment of four monitoring stations between Mathura and Agra.

iii. Cleaning of effluents emitted by the chimney of the Oil Refinery.

iv. Closing down of two thermal power stations at Agra.

v.   Developing a green belt of 1 to 5 km around the monument.

vi. Construction of an outer road to restrict traffic, as prevention is always better than cure.

Further, use of advanced technology should be made to save this magnificent historical monument, which is the pride of India.

OZONE

It is a highly reactive irritating gas with an unpleasant odour that forms in the troposphere as a major component of photochemical smog. Ozone is an important chemical species present in the stratosphere. At an altitude of about 30 km, its concentration is about 10 ppm. The ozone layer present in the stratosphere acts as a protective shield for the life on earth. It strongly absorbs ultraviolet radiations from the sun in the region 220–330 mm, allowing only a small fraction of UV radiation to reach the lower atmosphere of the earth's surface and thereby protects the life on earth from severe radiation damages such as DNA mutation and skin cancer.

Ozone Formation

Ozone is formed in the stratosphere by a photochemical reaction.

$$O_2 + hv \text{ (242 nm)} \longrightarrow O + O$$

$$O + O_2 + \text{(Third body such as } N_2 \text{ or } O_2) \longrightarrow O_3 + M$$

The third body (M) absorbs the excess energy liberalized by the above reaction and thereby the ozone molecule is stabilized. Thus ozone is constantly formed in the stratosphere.

Ozone Destruction

Ozone is also destroyed by the chlorine released due to volcanic activity and also by reaction with

i.   Nitric oxide

ii. Atomic oxygen

iii. Reactive hydroxyl radicals like HO_x, NO_x, ClO_x which are also present in the atmosphere by the following reactions.

$$O_3 + NO \longrightarrow NO_2 + O_2$$

$$O_3 + O \longrightarrow O_2 + O_2$$

$$O_3 + HO^\square \longrightarrow HO_2 + HOO^\square$$

$$HOO^\square + O \longrightarrow HO^\square + O_2$$

CFC and Ozone

Ozone in the stratosphere is also destroyed by man-made chloroflouro carbons (CFC), which are used as coolants in refrigerators, air conditioners,

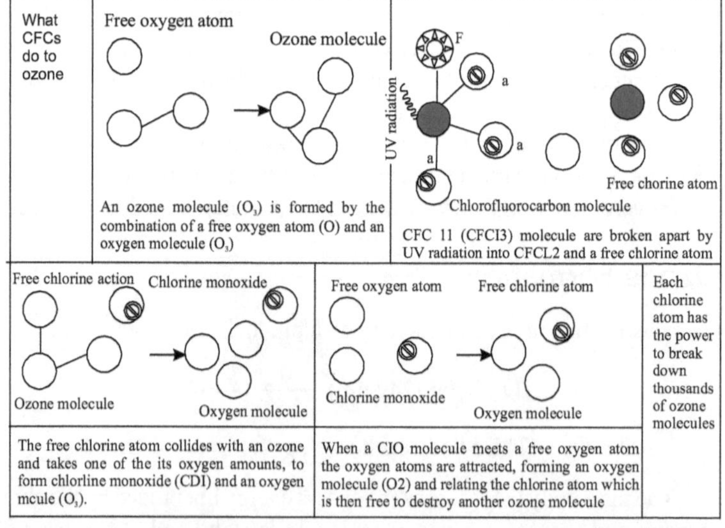

Figure 13.5   Breakdown of ozone molecules by CFCs

and propellants, in aerosol sprays, and in plastic forms. The CFC molecules liberated into the atmosphere reacts with ozone to liberate Cl atom. This liberated Cl is capable of attracting several ozone molecules as shown in Figure 13.5.

$$Cl + O_3 \longrightarrow ClO + O_2$$

$$ClO + O \longrightarrow Cl + O_2$$

This chain is involved which conserves clamorous hazards of CFC's were recognized in 1970. CFC is leading to the formation of ozone hole. Ozone hole was detected during September to November in1985. US immediately banned the use of CFCs in spray cans. Further in the year 1987, twenty-four nations of the world signed the Montreal protocol, which aims at 35% reduction in the global production of CFCs by year 1999. HFC-134 has been reported to be an effective substitute for CFC. Others like hydrofluoro carbons (HFCs), hydrochloro fluorocarbons (HCFCs) and methyl cyclohexane (MCH) are suggested for CFCs as substitutes.

Case Study 17

Ozone Hole Over Antarctica

The creation of ozone hole over Antarctica may be explained as follows: In the stratosphere, the CFCs are broken down by UV radiations and release chlorine atom. Using CFC-12, the reaction is

$$CF_2Cl_2 + hv \longrightarrow Cl + CF_2Cl$$

The chlorine atom will then react with ozone and produce CIO:

$$Cl + O_3 \longrightarrow ClO + O_3$$

The product ClO when reacts with NO_2, it forms chlorine nitrate:

$$ClO + NO_2 \longrightarrow ClONO_2$$

Chlorine nitrate, thus formed, is an inert compound that can do no damage to the ozone. Thus, at this state, Cl is effectively trapped in $ClONO_2$. Over Antarctica, however, a phenomenon of atmospheric circulation called the *circumpolar* or *polar vortex* forms. The formation of the circumpolar vortex blocks the warmer mid latitude air from mixing with the air above the pole. Thus the polar air is trapped with no connections to the outside warmer air.

This condition cools the air in the stratosphere, which can go down to $-90°C$. Even though the stratospheric air is very dry, still ice crystals can form at this very low temperature, providing reaction surfaces for chlorine nitrate to react with water to form HOCl and HNO_3:

$$ClONO_2 + H_2O \longrightarrow HOCl + HNO_3$$

As long as this polar vortex exists (during winter), the above reaction continues to operate, i.e., the accumulation of HOCl. This accumulation of HOCl is simply waiting for the Antarctic spring. As the sun first rises in the Antartic spring (of August or September), HOCl photolyses-forming Cl and the hydroxy peroxyl radical, which destroys the ozone. The reactions sequence is as under:

$$HOCl + hv \longrightarrow Cl + OH$$

$$Cl + O_3 \longrightarrow ClO + O_2$$

$$OH° + O_3 \longrightarrow HO_2^° + O_2$$

$$ClO + HO_2^° \longrightarrow HOCl + O_2$$

With this formation of HOCl, the cycle starts all over again. Each atom of chlorine chain then react s with thousands of molecules of ozone, thus converting ozone to oxygen.

Major human sources   It is obtained by chemical reaction with volatile organic compounds (VOCs) that are emitted mostly by cars and industries and nitrogen oxides to form photochemical smog.

Health effects   Breathing problems; coughing; eye, nose, and throat irritation; aggravates chronic diseases such as asthma, bronchitis emphysema, and heart disease; reduces resistance to colds and pneumonia; may speed up lung tissue aging.

Environmental effects   Ozone can damage plants and trees; smog can reduce visibility.

Property damage   Damages rubber, fabrics, and paints.

HYDROCARBONS

Natural biological and anthropogenic sources such as automobile exhaust; burning of coal, oil, wood and refuse will contribute hydrocarbons in air. 15% of total hydrocarbon is due to anthropogenic activities. The annual global hydrocarbon emission to the atmosphere by man-made source is roughly estimated to be 57×10^7 tons/per year.

Natural sources such as trees, emit large quantities of hydrocarbon in air, and plants like eucalyptus, cottonwood, oak, sweet gum and spruce trees emit isoprene. This isoprene undergoes several chemical changes to form particulate matter. Methane is emitted into atmosphere by anaerobic decomposition of organic matter by bacteria. In India, automobiles are the chief source of hydrocarbons. In Delhi, Bombay, and Bangalore, about 65% of hydrocarbons are emitted by automobiles. Unburnt hydrocarbons contribute about 40% of vehicular emission.

Sources	% of hydrocarbon
Petroleum	55.05
Gasoline	38.50
Refining	7.2
Oil	0.4
Evaporation of solvents in storage	8.8
Coal	3.3
Wood (fuel and fire)	2.2
Industrial use	0.8
Organic solvent evaporation	11.3
Incinerators	28.3
Power generation	0.2

These hydrocarbons being thermodynamically unstable tend to get oxidized in the atmosphere by a series of chemical and photochemical reactions. This gives rise to the formation of various end products such as CO_2, solid organic particulate, water-soluble acids and aldehydes which are washed down by rain.

Effects of Hydrocarbon

Photochemical reactions of hydrocarbons have more harmful effects than hydrocarbon itself. The photochemical smog which gives peculiar odour affects the vegetation, causes cracking of rubber products, eye irritation and reduced visibility.

Hydrocarbon in 100 ppm affects the mucous membrane secretion.

Methane is a severe gas pollutant. Its higher levels in the absence of oxygen create narcotic effects on human beings. A group of hydrocarbon causes cancer in man and animal, affecting DNA and cell growth.

Effects on Plants

Ethylene inhibits the plant growth, the damage of leaf tissues and the death of flowering plants.

Ethylene even at 1ppm concentration shows adverse effects on vegetations. Hydrocarbons and photochemical oxidants are injurious to plant. High exposure to ozone causes chlorosis.

Effects on Materials

Paper, textiles, rubber and polymers cannot withstand the changes inducted by ozone even at low levels.

LEAD

The solid toxic metal and its compounds are emitted into the atmosphere as particulate matter.

Major human sources   Paint (old houses), smelters (metal refineries), lead manufacture, storage batteries, leaded gasoline (being phased out in developed countries).

Health effects   Accumulates in the body; brain and other nervous system damage and mental retardation (especially in children); digestive and other health problems; some lead-containing chemicals cause cancer in test animals.

Environmental effects   Can harm wildlife.

CONTROLLING GASEOUS POLLUTANTS

Combustion

This technique is used when the pollutants contain gases or vapours, which are organic in nature. Finally, CO_2 is evolved when combustion of these pollutants in flame combustion or catalytic combustion where lower operating temperatures are desirable, e.g. burning of waste cracking gases, fumes from paint or enamel-baking ovens.

Absorption

In this technique, the gaseous effluents are passed through scrubbers or absorbers containing a suitable liquid absorbent to remove or modify one or more of the pollutants present in the gas stream. The gas absorption technique is widely used for removing pollutants like NO_x, H_2S, SO_2, SO_3 and fluorides from gaseous effluents. Some of the absorbents are H_2O, NaOH, ethanol, amines, soda ash, aluminum sulphate, etc.

Adsorption

The gaseous effluents are passed through porous solid adsorbents taken in suitable containers. The organic and inorganic constituents of effluent gases are held at the interface of the solid adsorbent by physical adsorption or chemisorptions. Adsorbents are silica gel, iron oxide, activated carbon, limestone, bauxite, etc. Description of the sorted gases is usually achieved by increasing the temperature or by reducing the pressure.

CONTROLLING PARTICULATE EMISSIONS

Gravity setting chamber, wet collection devices and electrostatic precipitators are used to remove dust.

Gravity Setting Chamber

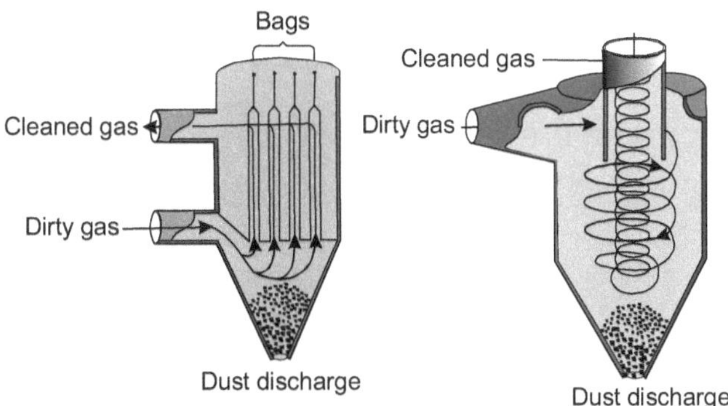

Figure 13.6   Cyclone separator

It consists of a chamber (Figure 13.6) in which dust is separated from gas by reducing the velocity of gas. As a result, dust particles (25–30 μm diameter) settle down in the chamber.

Wet Collection or Scrubbers

Particles are washed out of gas flow by a water spray. The object is to transfer suspended particulate matter in the gas to the scrubbing liquid, which can readily be removed by the gas-cleaning device, and ultimately gas is discharged to the atmosphere (Figure 13.7). Collected wastewater needs special setting tanks or filtration units.

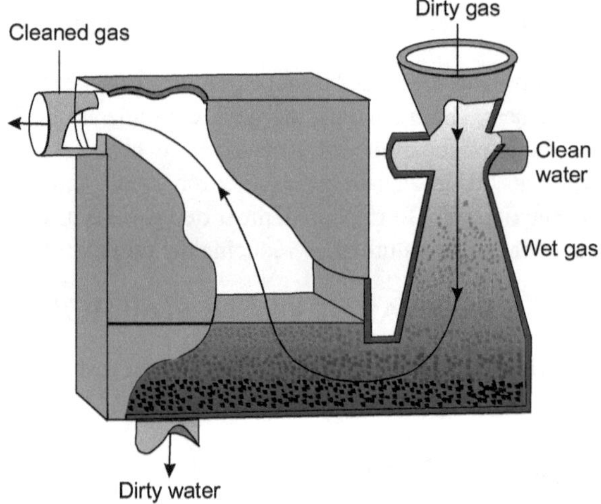

Figure 13.7   Wet scrubber

Electrostatic Precipitation

Small particles of diameters 0.0001 mm can be removed by passing the stack gas through an electrostatic precipitator. The electrostatic precipitator (Figure 13.8) consists of a series of plates, which are alternately charged to high voltages, and particles approaching the given plate tend to acquire charge. They are then attracted to the surface of the next plate, from which they fall into the hopper below. The efficiency is 99.9%. At 600°C about 1, 50,000 of gas/min is cleaned.

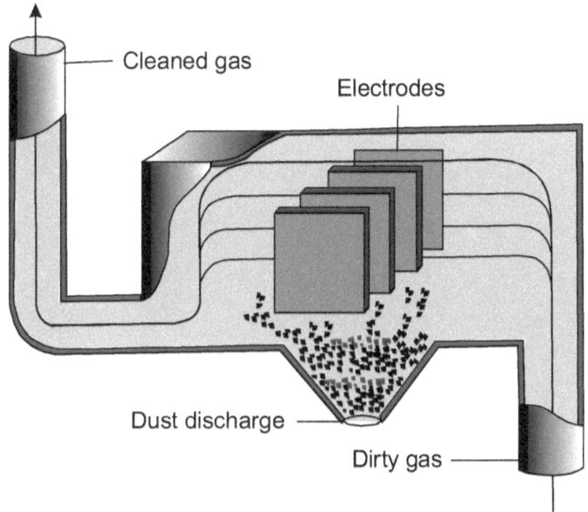

Figure 13.8   Electrostatic precipitator

Water is used for many purposes like drinking, domestic uses, industrial cooling, power generation, agriculture, transportation and waste disposal. In chemical industries, it is a solvent, scrubbing medium. Water resources like rivers, lakes, ponds are day-by-day getting polluted by many activities like, domestic, industrial, and agricultural and others (Figure 14.1). 70% of available water in India is polluted. From the Dal Lake in north to the Periyar River in the south and other rivers from east and west, water gets seriously polluted. Even the Ganga, our large perennial river is heavily polluted.

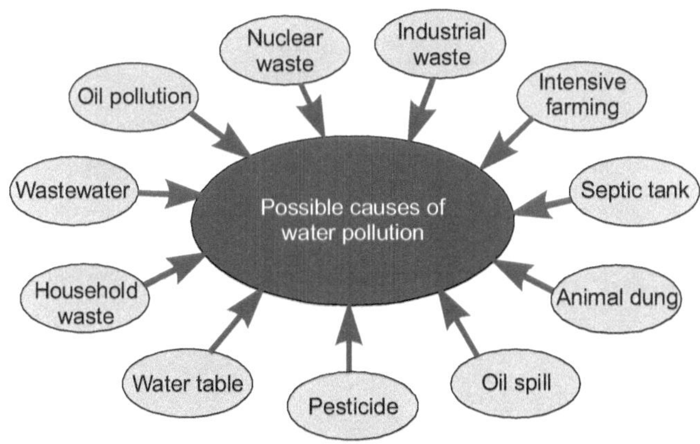

Figure 14.1 Various sources of water pollution

The implication of such massive pollution for the health of the nation is extremely serious. Two-thirds of all illnesses in India are related to water-borne diseases such as typhoid, infective hepatitis, cholera, diarrhoea and dysentery. In the environment, the pollutants may cause destruction of animal and plant life, and aesthetic nuisance. The major contaminants of drinking water are the pathogens as well as many toxic chemicals of industrial and agricultural activities. Thus, the waste treatment is of primary importance now.

Many of the rivers in urban centres are polluted. The aquatic life is disappearing because of the threat of mercury and pesticides in their body.

CAUSES OF WATER POLLUTION

There are many sources of pollution and the pollutants may be natural and artificial. Natural pollution is due to weather condition and rainfall, e.g. during storm, the banks of rivers are eroded and rivers are muddy carrying silt and vegetation (Figure 14.2).

Figure 14.2   A polluted river

Artificial Pollutants

Artificial pollutants are the sources, like domestic and industrial wastewaters. Pollution of the waterways is caused by any one or combination of many. Some of the important pollutants are the following

Oxygen-demanding compounds   The primary cause of oxygen demand of aquatic system is the presence of organic substances collectively called oxygen-demanding wastes. Although some inorganic substances are found in oxygen-demanding compounds, most of them are organic in nature, e.g. wastes generated from pulp and paper mills, tanning, food-processing industries, sewage and other organic wastes. These wastes decrease the dissolved oxygen level in water and produce odours, and impair domestic and livestock water supplies affecting taste, colour, and odour. Oils and detergents may protect water bodies from recreation of water bodies. Hot waters from industrial discharges lessen the dissolved oxygen level.

Artificial organic compounds   These are toxic poisons that persist for a long time from one tropical level to another and ultimately reach objectionable levels, e.g. pesticides, organic chemicals and detergents. Several pesticides have been shown to cause cancer in mice and rats. The detergents contribute approximately 50% to the phosphate present in sewage effluent. When these phosphates are released into streams and lakes, they act as plant nutrients, thus supporting eutrophic conditions.

Plant nutrients   Nitrogen and phosphorous are the essential elements required for the growth and metabolism of the plants and animals. These compounds may enter the water bodies directly from the manufacturer and use of fertilizers, and from the processing of biological materials such as food, textiles or via domestic sewage treatment plants. The concentration of these compounds in water bodies may lead to an excess growth of algae. This unsightly slime layer leads to eutrophication. When algae dies, this leads to subsequent increase in biodegradation which result in foul smell in waterbodies.

High concentration of nitrates in drinking water is also a major concern. When the nitrate-containing water is consumed, a complex with haemoglobin known as methaemoglobin is formed. As a result, the oxygen-carrying capacity of blood is reduced causing a condition of blue baby syndrome. Nitrate can be further converted to amines leading to gastric cancer in human body.

Inorganic chemicals and minerals   Inorganic salts, mineral acids, and finely divided metals and metal compounds are some examples of these pollutants in water. These pollutants enter the water bodies from municipal and industrial wastewaters, and mine runoff. Various metals and metallic compounds released from anthropogenic activities also add up to their natural background levels in water.

Some of the most toxic trace elements are the heavy metals, such as Hg, Cd, and Pb, and the metalloids such as As, Sb and Sc. The carboxylic acid (COOH) groups and amino (NH_2) groups of a protein may also be attacked by the heavy metal ions. The heavy metals that bound to the cell membrane interfere with the transport phenomena across the cell wall. They also tend to precipitate phosphate biocompounds or to catalyse their decomposition. Water pollution by heavy metals occurs mostly due to street dust, domestic sewage and industrial effluents.

Polyphosphates from detergents serve as algal nutrients and thus are significant as water pollutants.

Disease-causing wastes   These include pathogenic microorganisms that may enter the water along with sewage and other wastes and that may cause tremendous damage to public health. These microbes, comprising mainly of viruses and bacteria, can cause dangerous water-borne diseases such as cholera, typhoid, dysentery, polio and infectious hepatitis in human.

Sediments   Soil, sand, and mineral particles washed into aquatic environment by storms and floodwater are called sediments. Deposits of sewage sludge, pulverized coal, and various industrial solids are disposed off into rivers and marine waters. In suspension, the solids may cause

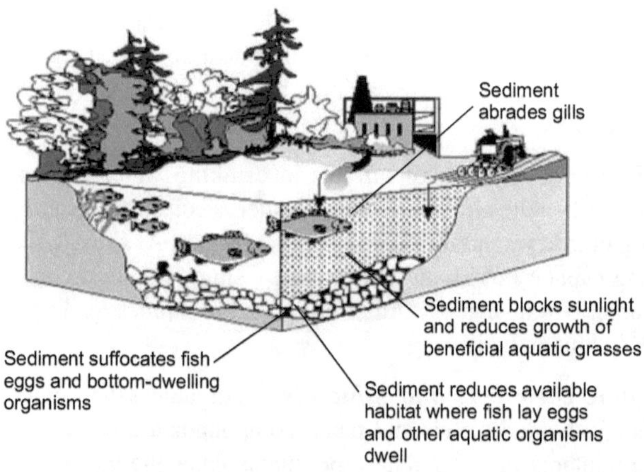

Figure 14.3   Siltation

thickening of fish gills, which may lead to eventual killing of fish. On the riverbed, sediments decrease the amount of food available for fish. Siltation is one of the leading pollution problems in the nation's river and streams (Figure 14.3).

Radioactive substances   The radioactive water pollutants may come from the following activities.

    i.   Mining and processing of ores

   ii.   Increasing use of radioactive isotopes

  iii.   Radioactive materials from nuclear power plants and nuclear reactors

The radioactive isotopes are toxic to life forms. The radioactivity that emanates from testing of nuclear weapons accumulates in bones and teeth and causes disorders in human beings.

Heat   Excess of heat is produced in all processes in which heat is converted into mechanical work. Thus, considerable thermal pollution results from thermal power plants. As heat-laden water is discharged back into the main water supply, the temperature of the aquatic environment is increased.

- Increase in water heat decreases the DO (dissolved oxygen) content of water.

- Reduction in DO in water may alter the spectrum of organisms that can adopt to live at their temperature and that DO level.

- It adversely affects the aquatic life.

- Suspended solids in water may also cause bad odours and tastes and also may promote conditions that are favourable for the growth of pathogenic bacteria.

Pollution from the production, transport and use of oil   The world's oil production is approximately three trillion gallons per year. Although less than 1% of this oil ends up as a pollutant, mostly in the oceans, this amounts to considerable contaminations. Thousands of oil spills contaminate coastal waters every year (Figure 14.4).

Figure 14.4   Oil spill clean-up (Source: www.umich.edu)

Oil spills also occur in inland waters. In January 1988, the collapse of an oil tank near Pittsburgh, Pennsylvania, sent million gallons of oil into the oceans. Losses of oil during off shore exploration, production of oil, accidental fires in ships and oil tankers, accidental or international oil slicks (as in the Gulf-war between Iraq and US led allied forces in the year 1991) and leakage from oil pipe lines, crossing waterways reservoirs. Oil pollution in sea has been increasing in recent years due to increase in oil-based technologies, massive oil shipments, accidental oil spillages and international oil slicks during international hostilities. The following are the effects of oil pollution.

- ◇   It causes the reduction of light transmission thereby reducing photosynthesis.

- ◇   It reduces the DO in water thus endangering water birds.

- ◇   It affects the coastal plants and animals.

Acids and alkalis   Acids and alkalis represent another kind of chemical pollution problem that has to do with ranges of tolerance for particular associates of living things in aquatic ecosystems. Only a few organisms can survive very high acidity. The processes such as battery manufacturing, electroplating, fertilizers, brewing, mining iron and copper, pickling, etc. discharge acidic wastes into the stream.

Chemical manufacturing wastes, kier liquors, wood scouring wastes, tannery wastes, cotton, and mercering wastes are the principal contributors of alkalis. Fish and other aquatic life are also affected by a sudden rise in pH.

Table 14.1 lists out the important pollutants of water and their effects.

Table 14.1   Pollutants of water and their effects

Pollutants	Effects
Oxygen-demanding wastes	Decrease DO; change the odour, colour, and taste of water.
Artificial organic compounds	Produce cancer in mice and rats; discharge nutrients in aquatic system leads to eutrophication.
Plant nutrients	Cause eutrophication, which on biodegradation leads to fowl smell.
Inorganic chemicals and minerals	Heavy metals interfere with transport of cell water, precipitate phosphate biocompounds, and phosphate leads to algal nutrients.
Disease-causing wastes	Cholera, typhoid, dysentry, polio, infectious hepatitis.
Sediments	Decrease the amount of food available for aquatic life.
Radioactive wastes	Isotopes are toxic to life forms accumulates in bones and teeth causing disorders in human beings.
Heat	Decrease the DO content; Reduction of DO decreases aquatic life.
Oil	Reduction in light transmission; coastal plants and animals affected.
Acids and alkalis	Affect fish and other aquatic life.

TYPES OF WATER POLLUTANTS

According to the American College Dictionary, pollution is defined as: "to make foul or unclean; dirty." Water pollution occurs when a water body is adversely affected due to the addition of large amounts of materials to the water. When it is unfit for its intended use, water is considered as polluted.

Two types of water pollutants exist: point sources and non-point sources. Point sources of pollution occur when harmful substances are emitted directly into a body of water. The Exxon Valdez oil spill best illustrates a point source water pollution. A non-point source delivers pollutants indirectly through environmental changes. An example of this type of water pollution is when fertilizer from a field is carried into a stream by rain, in the form of run-off which in turn affects aquatic life. The technology exists for point sources of pollution to be monitored and regulated, although political factors may complicate matters. Non-point sources are much more difficult to control. Pollution arising from non-point sources accounts for a majority of the contaminants in streams and lakes (Figure 14.5).

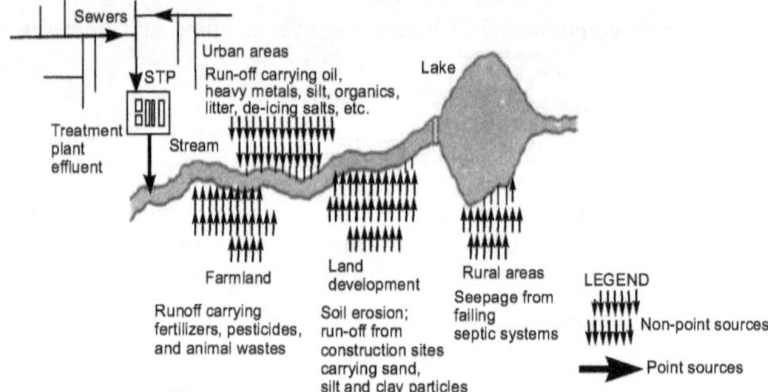

Figure 14.5   Source pollutants

CHARACTERISTICS OF WASTEWATERS

Wastewaters are characterized on the basis of various physical, chemical and biological characteristics.

Type	Characteristics
Physical characteristics	Colour, odour, dissolved oxygen (DO), insoluble substances (settable solids, suspended solids), corrosive properties, radioactivity, temperature range, and formability
Chemical characteristics	Chemical oxygen demand (COD), pH, acidity or alkalinity, hardness, total carbon, total dissolved solids, chlorine demand, known organic and inorganic components such as Cl, SO_2, SO_4, N, P, Pb, Cd, Mg, C_7, surfactants, phenols, hydrocarbons, oils and greases.
Biochemical characteristics	Biological oxygen demand (BOD), i.e., presence of pathogenic bacteria, etc., which are harmful to human, plants and animals.

SELF-PURIFICATION OF POLLUTED WATERS

Water acts as sink for all the wastewater. At the point of outfall, heavier suspended solids settle down and the portion of water ahead becomes free of them. The remaining of the suspended and dissolved solids gets dissolved into the large volume of receiving water and gets diluted. The voltaic organic material is a source of food for bacteria and other organisms who decompose and stablize the same. The action will be aerobic when sufficient DO is available and anaerobic in the absence of oxygen. The mineralized materials act as a food to algal and other microscopic plant life that thrive and supply oxygen to the stream for maintaining aerobic actions. The protozoans and other animals live upon the bacteria while they themselves will fall prey to the fish and other aquatic life. The settled organic solids get decomposed slowly in an anaerobic way and will get ultimately mineralized and gasified. They are finally washed out by the flowing stream. The receiving water thus becomes free of pollution and will ultimately go to its normal state within a certain distance or time. Simultaneously there is replenishment of oxygen through the surface, at a rate that is proportional to the depletion of oxygen below the saturation value. The simultaneous action of deoxygenation and reaction produces a typical pattern in the dissolved oxygen concentration of the aquatic system. This pattern is known as the dissolved oxygen sag and a typical curve is shown in Figure 14.6.

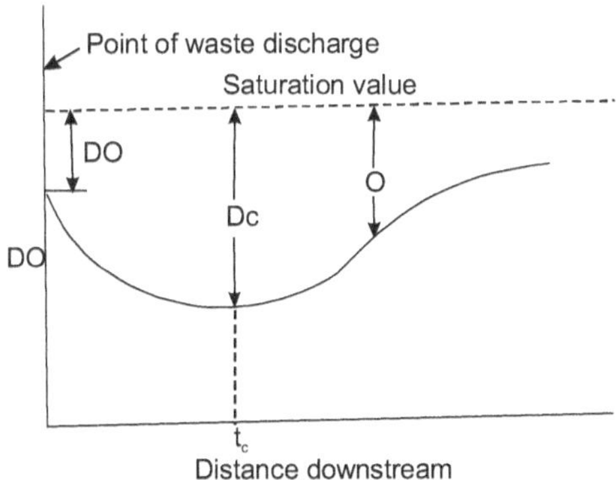

Figure 14.6   Oxygen sag curve

The sag curve initially drops as the wastes deplete the oxygen faster than it is replenished. At the point where the dissolved oxygen (DO) is a minimum, rate of reaeration becomes equal to the rate of deoxygenation.

Beyond this point, the rate of reaeration exceeds the rate of deoxygenation and the level begins to increase and eventually returns to normal.

WASTE QUALITY STANDARDS

The analysis required for water samples depend on the intended use of the water. For example, if its intended use is drinking, water should meet certain quality criteria with respect to the appearance (turbidity, colour, etc.), portability (tastes, odour, etc.), health (baceria, nitrates, chlorides, etc.) and toxicity (metals, organics, etc.). These and similar criteria are established by health or other regulatory agencies (Indian Council of Medical Research (ICMR), World Health Organization (WHO), United States Public Health Services (USPHS), etc.) to ensure that the water quality in a resource is suitable for the proposed use.

Based on the criteria, quality standards are set, which reflect the current state of knowledge of various constituents. The standards are continuously revised, as more and more is learnt about the effluents of different usage. Hence, these standards should not be used as absolute limits, but only as guidelines that can be used for preliminary judgements.

CONTROL MEASURES

WASTEWATER TREATMENT

The purpose of wastewater treatment is to remove the contaminants from water so that the treated water can meet the acceptable quality standards. The quality standards usually depend upon whether water will be reused or discharged into a receiving stream (Figure 14.7).

Principles of Wastewater Treatment

Most biological waste and wastewater treatment processes employ bacteria as primary microorganisms and certain other microorganisms may also play an important role. Characteristics of wastewater are measured in terms of chemical oxygen demand (COD), biochemical oxygen demand (BOD) and volatile suspended solids (VSS). Degeneration of organic matter is effected by its use as food by microorganisms to provide protoplasm for new cells during the growth process.

pH, temperature, type and concentration of the substrate, hydrogen acceptor; essential nutrient concentration and availability; concentration of essential nutrients (e.g. nitrogen, phosphorous, sulphur, etc.); and essential minerals are the environmental factors.

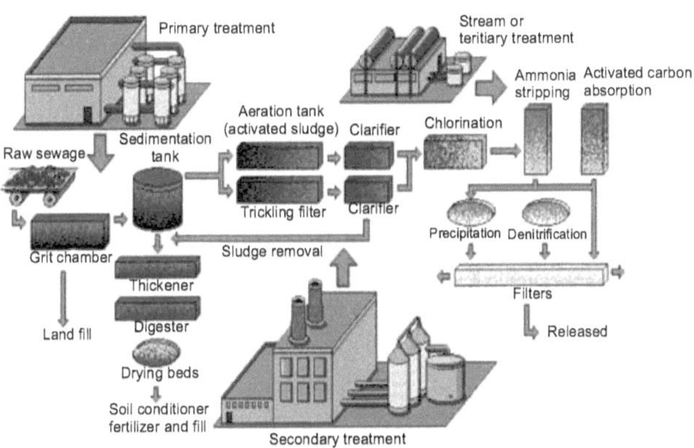

Figure 14.7   Wastewater treatment

Metabolic Reactions

Metabolic reactions occurring within a biological treatment process can be divided into three phases .

1. Organic matter oxidation (respiration)

$$C_x H_y O_2 + O_2 \longrightarrow CO_2 + H_2O + energy$$

2. Cell material synthesis

$$C_6 H_6 O_3 + NH_3 + O_2 \longrightarrow C_5H_7NO_2 + CO_2 + H_2O$$

3. Cell material oxidation

$$C_5H_7NO_2 \longrightarrow NH_3 + 5CO_2 + 2H_2O + energy$$

Oxidation–reduction   Oxidation–reduction reactions continue either in the presence of free oxygen (aerobically) or in its absence (anaerobically). The overall reaction may be different under aerobic/ anaerobic conditions but the microbial growth and energy utilization are similar.

Primary Treatment

The principle objectives of primary treatment are the removal of gross solids (i.e., large floating and suspended solid matters, grit, oil and grease) if they are present in considerable quantities.

Large quantities of floating rubbish such as cans, cloth, wood and other larger objects present in the wastewater are usually removed by metal bars. They act like strainers, as the wastewater moves beneath them in an open channel. The velocity of the water is then reduced in a grit-settling chamber of a larger size than the previous channel.

The modern mechanical screens cum filters include rotary, self-cleaning, gravity type units and circular overhead fed vibratory units. These are costlier, as compared to the conventional bar screens, but are very effective in reducing the suspended solids and BOD. Sometimes, instead of screening the gross solids in the sewage, they are cut into small pieces with the help of macerators or comminutors.

The grit (or detritus) is removed in the early stages of treatment in grit channels or tanks to safeguard against any damage to pumps and other equipment by abrasion and also to avoid settling in pipe bends and channels. Grits, being heavier than organic solids, can be separated from organic solids by careful regulation or the flow velocity in the grit tanks.

If the oil and grease are in emulsified condition, as in wool-scouring wastes, ordinary skimming methods are ineffective. In such cases, they may be removed with the help of chemical reagents in primary sedimentation tanks.

After the removal of gross solids, gritty materials and excessive quantities of soil and grease, the next step is to remove the remaining suspended solids as much as possible. This step is aimed at reducing the strength of the waste-water and also to facilitate secondary treatment.

Sedimentation   The suspended matter can be removed effectively and economically by sedimentation. This process is particularly useful for treatment of wastes containing high percentage of settleable solids or when the waste is subjected to combined treatment with sewage.

The sedimentation tanks are designed to enable smaller and lighter particles to settle under gravity. The most common equipment used included horizontal flow sedimentation tanks and centre feed circular clarifier. The settled sludge is removed from the sedimentation tanks by mechanical scrapping not hopper and pumping it out subsequently. In a well-designed continuous flow sedimentation tank, about 50% of the suspended solid matter is settled out within two hours of detention time. An efficient sedimentation system is expected to remove about 90% of the suspended solids and 40% of the organic matter (thus reducing the BOD).

In industries having large proportions of wastewater, long detention time helps in the mixing and equalization of various wastes before being sent to the biological purification plants.

Sedimentation aids Finely divided suspended solids and colloidal particles cannot be efficiently removed by simple sedimentation by gravity. In such cases, mechanical flocculation or chemical coagulation is employed.

In mechanical flocculation, the wastewater is passed through a tank with a detention time of 30 minutes and fitted with the paddles rotating at an optimum peripheral speed of 0.43 m/s. Under this gentle stirring, the finely divided suspended solids coalesce into larger particles and settle out. Specialized equipment as Clariflocculator is also available, wherein flocculating chamber is a part of sedimentation tank.

Coagulation is the most effective and economical means to remove impurities.

The sequence of operations in the chemically aided coagulation is

i.   Addition of lime, if there is no sufficient alkalinity

ii. Addition of coagulant, followed by rapid mixing for 4 to 6 minutes

iii. Addition of coagulant aids, followed by gentle agitation or slow stirring for about 40 minutes.

Equalization Some industries produce different types of wastes, having different characteristics at different intervals of time. Hence, uniform treatment is not possible. In order to obviate this problem, different streams of effluents are held in big holding tanks for specified periods of time. Each unit volume of waste is mixed thoroughly with other unit volumes of wastes to produce a homogeneous and equalized effluent. Aeration or mechanical agitation with paddles usually results in better mixing of the different unit volumes of effluents.

Neutralization   Highly acidic or highly alkaline wastes should be properly neutralized before discharged into the stream. Acidic wastes are usually neutralized by treatment with lime slurry or caustic soda, depending upon the type and quantity of the waste. Alkaline wastes may be neutralized by treatment with sulphuric acid or CO_2 or waste boiler fuel gas.

Secondary Treatment

In the secondary treatment, the dissolved and colloidal organic matters present in the wastewater are removed by biological processes involving bacteria and other microorganisms. These processes may be aerobic or anaerobic. In aerobic processes, bacteria and other microorganisms consume organic matter as food. They bring about the following sequential changes.

   i.   Coagulation and flocculation of colloidal matter

   ii. Oxidation of dissolved organic matter to CO_2

   iii. Degradation of nitrogenous organic matter to ammonia, which is then converted into nitrite and eventually to nitrate

Thus, secondary treatment reduces BOD. It also removes appreciable amounts of oil and phenol. However, commissioning and maintenance of secondary treatment system are expensive.

The effluent from primary sedimentation tanks is first subjected to aerobic oxidation in systems, such as aerated lagoons, trickling filters, activated sludge units, oxidation ditches or oxidation ponds. Then the sludge obtained in these aerobic processes, together with that obtained in the primary sedimentation tanks, is subjected to anaerobic digestion in the sludge digesters.

Certain microorganisms, in the presence of dissolved oxygen and in proper environmental conditions, utilize organic wastes as their foods, and convert them into simpler compounds, such as CO_2, nitrates and sulphates, which are non-pollutants. This process therefore can be used to remove organic substances from wastes. Almost all organic substances, with a few exceptions, such as hydrocarbons and others, can be oxidized by aerobic biological treatment. Complex cell tissues and protein materials are also synthesized during this process, which are then agglomerated and removed from the wastes by settling. Germicidal and resistant organic compounds such as cyanides and phenols, can also be destroyed by special types of microorganisms, after prolonged acclimatization periods.

Under anaerobic conditions (i.e., in the absence of dissolved oxygen or gaseous oxygen), certain groups of microorganisms, e.g. hydrolyte and methane-forming organisms, can carry out the digestion of complex organic wastes. The hydrolyte organisms convert complex organic compounds to simple and low-molecular-weight organic acids and alcohols. These are then converted to CO_2 and CH_4 by methane bacteria. Anaerobic treatment processes can be carried out in-depth without the need for large surface area. It can take place in mixed or enriched cultures and can, therefore, be maintained easily on a large scale. This process can be applied to most types of substrates excepting a few like lignin and mineral oil. The process is less-expensive but the final effluent is less satisfactory, as compared to that from aerobic treatment, because of the dark colour, odour and higher residual BOD.

Anaerobic treatment is mainly employed for the digestion of sludges. However, organic liquid wastes from dairy, slaughterhouse, etc., were economically and effectively treated by this method. The efficiency of this process depends upon the pH, temperature, waste loading, absence of oxygen, and toxic materials.

Some of the commonly used biological treatment processes are described below.

Aerated lagoons   These are large holding tanks or ponds having a depth of 3–5 m and are lined with cement, polythene or rubber. The effluents from primary treatment processes are collected in these tanks and are aerated with mechanical devices, such as floating aerators, for about 2 to 6 days. During this time, a healthy flocculent sludge is formed which brings about oxidation of the dissolved organic matter. BOD removal to the extent of 90% could be achieved with efficient operation. The operation and maintenance are relatively simple. The major disadvantages are the larger space requirement and the bacterial contamination of the lagoon effluent, which necessitates further the biological purification in maturation pond or by secondary sedimentation and sludge digestion.

Trickling filters   The trickling filters usually consist of circular or rectangular beds, 1 to 3 m deep, made of well-graded media (such as broken stone, PVC coal, coke, synthetic resins, gravel or clinkers) of size 40 to 150 mm, over which the wastewater is sprinkled uniformly on the entire bed with the help of a slow rotating distributor (such as rotary sprinkler equipped with orifices or nozzles). Thus, the wastewater trickles through the media. The filter is arranged in such a way that air can enter at the bottom, counter current to the effluent flow and a natural draft is produced. A gelatinous film, comprising of bacteria and a aerobic microorganisms known as "Zooglea" is formed on the surface of the filter medium, which thrive on the nutrient supplied by the sewage or the wastewater. The organic impurities in the wastewater are adsorbed on the gelatinous film during its passage and then are oxidized by the bacteria and the other microorganisms present therein. When the thickness of the film on the medium increases, a part of it gets detached and carried along the effluent. Hence, the effluent from the trickling filters is allowed to settle in a settling tank to retain the sludge particles and is then discharged. The sludge is then pumped to the sludge digestion unit.

The microbial film formed is very sensitive to temperature. The metabolic activity is proportional to the temperature of the wastewater passing through the filter. Thus, the efficiency of the matter decreases in winter season. The efficiency of the filter depends upon the composition of the waste, strength of hydraulic loading, temperature, pH, depth of the filter, the size and uniformity of the filter medium, uniformity of wastewater distribution over the filter and proper air supply.

Trickling filters are simple to operate and can produce BOD removal to the extent of 65 to 85% depending upon the rate of filtration. Moreover, constant manual attention is not needed for this process. Trickling filters produce effluents of consistent and better quality.

Trickling filters are effectively used for the treatment of industrial wastes from dairy, distillery, brewery, cannery, food processing, pulp and paper mills, pharmaceuticals, petrochemicals, slaughterhouse and poultry processing industries.

Activated sludge process   This is the most versatile biological oxidation method employed for the treatment of wastewater containing dissolved solids, colloids and coarse solid organic matter. In this process, the sewage or industrial wastewater is aerated in a reaction tank, in which some microbial floc is suspended. The aerobic bacterial flora brings about biological degradation of the waste into SO_2 and H_2O, while consuming some organic matter for synthesizing bacteria. The bacterial flora grows and remains suspended in the form of a floc, which is called "activated sludge". The effluent from the reaction tank is separated from the sludge by settling and is discharged. A part of the sludge is recycled to the same tank to provide an effective microbial population for a fresh treatment cycle. The surplus sludge is digested in a sludge digester, along with the primary sludge obtained from primary sedimentation. An efficient aeration for 3 to 6 hours is adequate for sewage, whereas for industrial wastes, 6 to 24 hours of aeration is required for this process. BOD removal to the extent of 90–95% can be achieved.

For this process to be efficient, at least 0.5 ppm oxygen must be present all the time. Oxygen is supplied either by mechanical aeration or by diffused aeration system.

The microorganisms should be provided with essential nutrients such as N and P which are supplied in the form of urea and mono ammonium hydrogen phosphate. Other nutrients, e.g. K, Mg, Ca, etc., are generally present in the wastes. Other important factors that determine the efficiency of the activated sludge are pH, temperature and oxidation–reduction potential. The optimum pH range for the process is 6.5 to 9.0. Low temperature slows down the rate of metabolism, while high temperature increases the metabolic activity to such an extent that the oxygen is consumed fast, leading to anaerobic conditions.

The performance of the activated sludge process can be assessed with the help of microbial indicators, as observed by regular microscopic examination of the activated sludge. A good activated sludge contains a relatively high population of free swimming (e.g. stylonichia) and stalked ciliates (e.g. vorticella) apart from a few rotifers. A very high population of dispersed bacteria indicates a poor activated sludge system. The presence of filamentous microorganisms indicates the deficiency of N or P, low pH and low oxygen levels. The growth of filamentous bacteria may retard the floc compaction and settling and result in turbid effluents.

Although anaerobic treatment is a slow process, it is useful for treating small quantities of wastes, containing readily oxidizable dissolved organic solids in liquid form or in finely divided form. The operation and maintenance costs are lesser with this treatment. That is why some liquid wastes containing soluble organics from the dairy, slaughterhouse and paper mill industries have been economically and effectively treated by this process.

About 0.6 to 1.25 m^3 of the gas is produced per kg of the organic matter destroyed. The calorific value of the gas is about 26 MJ/m^3. The gas can therefore be used as a fuel to provide the heat required to warm the digestion tanks. In large installations, it can be used for power generation.

Sludge treatment and disposal   The sludge from the digester may contain about 90 to 98% water. The sludge is dewatered in drying beds, filter process or vacuum filters. The dewatered sludge, after chlorination, can be sent for ultimate disposal. The various methods used for ultimate disposal include dumping in landfills, incineration, dumping at selected sites in sea, or utilizing as a low-grade fertilizer.

Tertiary Treatment

Tertiary treatment is the final treatment, meant for polishing the effluent from the secondary treatment processes, to improve its quality further. The major objectives of tertiary treatment are:

    i.   removal of fine suspended solids

    ii. removal of bacteria

    iii. removal of dissolved inorganic solids

    iv. removal of final traces of organics, if it is felt necessary

Removal of finely divided suspended solids can be achieved with the help of micro-strainers and sand filters.

Removal of bacteria, particularly of faecal origin, can be achieved by retaining the effluent from secondary biological treatment plants in maturation ponds or lagoons for specified periods of time. Three or four lagoons arranged in series give an excellent final effluent with very low BOD and low suspended solids. The final effluent is chlorinated if necessary.

Removal of dissolved inorganic solids is a major problem with wastewaters from industries such as fertilizers, textiles processing, tannery and electroplating. Depending upon the required quality of the final effluent and the cost of the treatment that can be afforded in a given situation, any of the following treatment methods can be employed.

Evaporation   This is an energy-intensive and hence an expensive process. It is used only when the recovered solids or the concentrated solutions are used, e.g. some electroplating wastes and when the volume of the wastewater to be treated is less (this method used). This method is also employed for concentrating radioactive liquid wastes. It permits the recovery of wide varieties of processing chemicals. It is applicable to remove or concentrate chemicals, which cannot be accomplished by any other means.

Ion-exchange   The use of ion-exchange for demineralization of water is well known. It is widely used for obtaining deionized water for use in high-pressure boilers. This process is now extended to wastewater treatment for the removal and recovery of toxic materials from wastewater. Ion-exchange process is economical only when the recovered salts are reused in the process, as in electroplating industry. Despite the simplicity of its operation, the method may not be economical if the objective of the treatment is only the removal of dissolved solids from wastewater. Special ion exchange is available for the retrieval of toxic metal ions from industrial wastewater. Applications of ion-exchange process in wastewater treatment include the recovery of Cr, Ni, phosphate and H_2SO_4 from anodizing baths, the recovery of Cu, Pb and Hg, recovery of Cr from cooling tower below down and the removal of cyanides from waste streams.

Adsorption   It is a process that occurs when a gas or liquid solute accumulates on the surface of a solid or a liquid (adsorbent), focusing a film of molecules or atoms (the adsorbate). Adsorption by activated carbon is advantageous to remove small quantities of organic contaminants from wastewater. Special adsorbents are commercially available for the removal and retrieval of toxic heavy metal ions from industrial wastewater. Activated carbon treatment is particularly useful for the removal of pesticides (e.g. dichloro-diphenyl trichloroethane) and carbonate insecticides.

Adsorption processes using activated carbon, peat moss, brown coal and other cellulose materials are finding increasing application in wastewater treatment for the removal of refractory, organic, toxic metals, colour, etc.

The ALM series of adsorbents developed in Japan incorporate high polymer products containing S and N functional groups. The products have very high affinity for heavy metals and can reduce metal levels in mining and industrial wastewaters to less than 1 ppb.

Electrolysis   This is a process in which the colloidal/dissolved species are exchanged between two liquids through selective ion-exchange membranes. An electromotive force brings about the separation of the species according to their charge. The semi-permeable membranes allow the passage of certain charged species while rejecting the passage of oppositely charged species.

Applications of this process include:

i. concentration of rinse waters to the desired bath strength
ii. removal of Cu, Zn, Ni, chromic acid, fluoride and cyanide from rrocess waters
iii. recovery of Cr from automobile plating rinse baths
iv. recovery of valuable metals and radioactive elements
v. control of water pollution
vi. In desalination of water
vii. In purification of plasma protein
viii. In pharmaceutical, medical and photographic industries
ix. In demineralization of sugars

Electrolytic recovery   In this process, electrochemical reduction of metal ions to elemental metal takes place at the cathode. This process is used to recover Cu, Sn, Ag and other metals from plating, pickling baths. The efficiency of this process has been tremendously increased by innovative designs such as eco cell involving rotating electrodes (particularly suitable for concentrated wastes) and extended surface electrolysis (specially suitable for dilute wastes).

Reverse osmosis   This is essentially a process in which a semi-permeable membrane allows water molecules to pass through and retains the ions. A thin cellophane sheet supported on a filter cloth can be used as a successful membrane. High pressures are required for useful flow rates, and recovery can be achieved by back flushing. High-pressure pumps of the order of 500 psi are required.

The various designs of reverse osmosis equipment include flat plate, tubular, spiral or jelly roll and hollow fibre types.

Applications of reverse osmosis include recovery of valuable components from effluents, recovery of water for reuse, pollution recycling of wastewaters and spent chemicals from metal plating industry.

SOIL

Soil is one of the most significant ecological factors, which is derived from the transformation of surface rocks. It is nothing but the soil on which plants depend for their nutrients, water and mineral supply and anchorage. It forms an important medium wherein numerous animals live. It is a valuable material heritage. Soil provides homes and ideal environmental conditions for living beings.

Life on earth depends directly on the living soil and aquatic ecosystem of rivers. Only in fertile soil, microbial fauna habit in. Plants will grow and nutrients will be recycled. The soil flora and fauna are affected when the earth soils are being stripped away, contaminated with toxic chemicals at a rate that cannot be sustained.

Most soil area used for agriculture either lack water or have physical and chemical constraints like steep slopes, easily eroded or poorly drained soils, alkalinity, salinity and other conditions that are toxic to plants. At present, the world's cropland averages about 0.28 hectare per capita. If the world's food production is to increase 60% by the year 2025, as it is must to maintain current nutrient levels, their croplands must expand or crop yields must increase. The latter require more inputs of fertilizers, pesticides and irrigation, with the risk of worsening soil pollution.

SOIL STRUCTURE

Vertical Distribution of Soil

Soil is composed of distinct layers such as topsoil, subsoils, weathered rocks, and bottom bedrock (Figure 15.1 and 15.2). Topsoil is composed of dead organic matter of plants and animals, which are reduced to fine particles. Roots of small plants are embedded in topsoil. It is a leaching zone. Below the topsoil is the subsoil which is coarse textured. Soil texture may be sandy soil, clay soil, loam soils, or may be sandy loam soil, clay loam soil, or silt loam soil. The bottom most is the rock where water is collected.

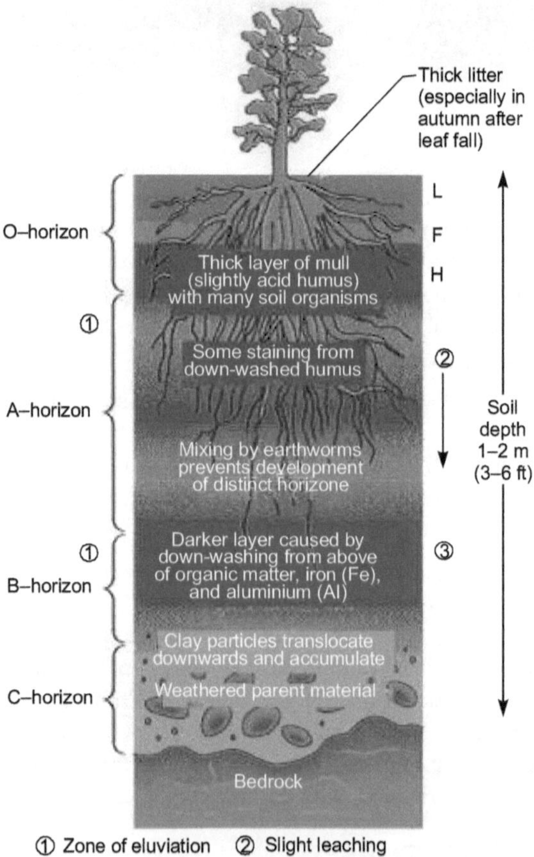

Figure 15.1 Vertical distribution of soil

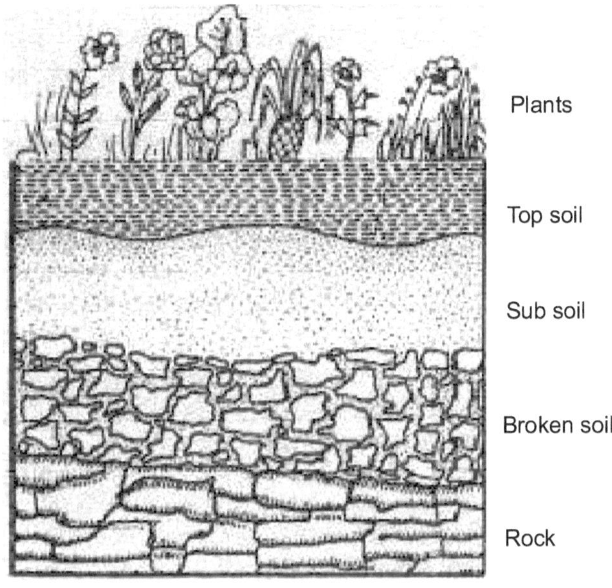

Figure 15.2   Vertical distribution of soil

SOIL TYPE AND PLANT INDICATORS

Forests serve as good indicators of soil productivity. Growth of certain trees like *Stellata* is poor in sterile sandy soil. *Polygonum* is a short-lived perennial which grow better in over-grazed areas. Grasses like *Saccharum spontaneum* prefer to grow in poorly drained soils. Plants like *Anabasis aphylla* can grow and tolerate sulphate concentration of the soil. Salinity of soil is reflected by the absence of salt-sensitive plants. Chloride salinity of soils enhances the growth of *Kalidium capsicum*. Some plants are calciphytes, which grow on basic soils (Figure 15.3). Certain plants can withstand the presence of heavy metals in higher concentration, e.g. silene plant to zinc, liver worts and mosses to copper, serpentina to chromium, and acidophilic grass to aluminum. Presence of diatoms indicates sewage pollution, while bacteria such as *E. coli* indicates water pollution. Soil pollution affects the physiological process inhibiting plant photosynthesis causing bleaching of stamen.

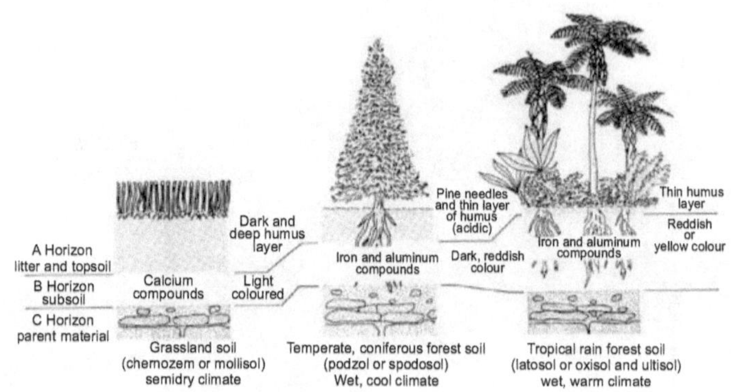

Figure 15.3   Soil profiles of various plants

SOURCES OF SOIL POLLUTION

Nowadays there is a rapid industrialization and urbanization and rapid advanced technology. The impact of human is more in all the natural sources like land, water, air, etc. Every year solid wastes are increasing tremendously. Several hazardous chemicals and solid wastes are dumped on soil. Modern agricultural practices introduce numerous pesticides, fungicides, bactericides, insecticides, biocides, fertilizers and manures, resulting in severe biological and chemical contamination of land. Rubbish from construction activities, dead animals and discarded manufactured products lead to huge amount of wastes which causes serious disposal problems.

The intensity of waste problems in land lies in the leachates and mounting amounts of wastes. The leachates moves down the soil and pollute the subsoil as well as the groundwater. Unlike air pollution, land pollution remains in direct contact with soil for relatively longer periods. Thus the soil is getting heavily polluted day-by-day by toxic materials, which enter the air, water and food chain. The nature of pollutants in the soil is given in Table 15.1. The following are the few sources of land or soil pollution.

1. Domestic wastes and commercial wastes
2. Industrial wastes
3. Hazardous wastes
4. Agricultural wastes

5. Chemical wastes

6. Biological wastes

Table 15.1 The nature of pollutant in the soil

Source	Gases	Colloids	Suspended particles
Soil	CO_2	Clay, Fe_2O_3 Al_2O_3, MnO_2	Clay, sand, silt
Decomposed organic matter	SO_2, H_2, NH_3, CH_4, CO_2	Organic waste materials	Human organic wastes
Soil organism	-	Algae, fungi, bacteria, protozoa, viruses	Algae, bacteria

Domestic and Commercial Wastes

Solid wastes and refuses particularly in urban areas contribute soil pollution. About 135 million populations are living in urban areas in India. Important cities like Bombay, Calcutta, Kanpur, and Chennai are contributing 5,000 tons of waste material in a day. The leachates from dump sites and disposal of sewage mixed with industrial effluents are toxic to soil and the habitats. When people dump their waste products into the soil, then the leachate generated poses problem to groundwater.

Effects of wastewater on soil   The letting of wastewater especially from domestic sewage into the soil contaminates the groundwater and sub surface water. The other causes are detailed below:

1. A direct public health hazard

2. External groundwater pollution

3. Accumulation of hazardous substances in the soil or water, which can enter into the food chain

4. An account of pollutants such as odours into the atmosphere

5. Aesthetic loses within the limits

Effects of solid wastes on soils   The contaminants while on leachates join the aquifers and cause danger to groundwater and surface water. Though the organic matter present in wastes increases the soil productivity, excess is extremely dangerous. Solid wastes mainly constitute animal manures, food materials, sewage sludges, municipal wastes, leaves and garbage composts.

Industrial Wastes

Industrial wastes from paper, distilleries, oil refineries, tanneries, textiles, metal processing, mining, agro-industries and others are estimated to produce 50% of their raw materials as waste (Figure 15.4). What is happening to these wastes? These wastes pollute, affect and alter the chemical and biological properties of soil. Tannery wastes contain the pathogenic bacteria, which affect animals and plants.

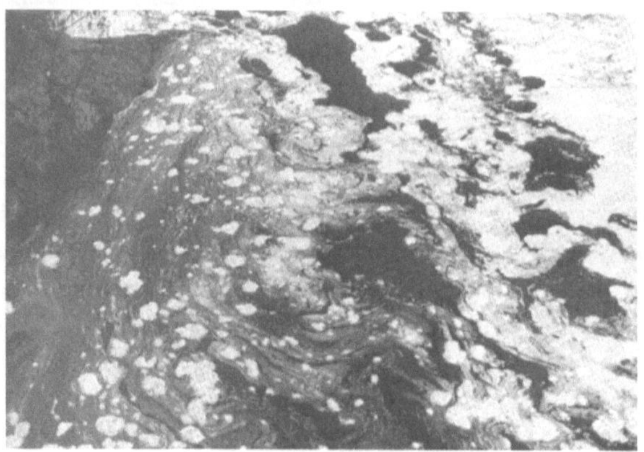

Figure 15.4   Waste from the paper mill

Agricultural Wastes

Farmhouse wastes, phosphates, nitrates, etc. mix with water present in the soil. When these waters reach the lake or nearby rivers, they create various health hazards and eutrophication. Some of the reasons responsible for decreased food production and cultivable areas are as follows:

i.   The cost of intensive agriculture has often resulted in soil erosion and depletion of groundwater resources.

ii. Each year the requirement of more fertilizers, chemical pesticides and farm machinery is needed in cropped areas to produce the same quantity of grain.

   iii. Damage by pest, insects and disease got worse.

   iv. Modern agronomy practices enhanced the problem of accumulation of greenhouse gases in the troposphere.

Alternative agricultural practices   In order to eliminate the above setbacks in the agriculture sector, the following alternative agricultural practices may be adopted.

   i. Reducing the need of off-farm inputs and chemicals.

   ii. Make greater use of biological and genetic potential of plant and animal species.

   iii. Incorporate natural processes such as nutrient cycles, nitrogen fixation in soil, etc.

   iv. Emphasize improved farm management and conservation of soil, water energy and biological resources.

Chemical Wastes

Textile, pesticides, paints, dyes, soap, pats, paper, sugar, electroplating discharge hazardous effluents in soil and create effects on living organism. They also discharge metallic contaminants in soil, which are considered to be poisonous and accumulate in plants and in turn affect living beings.

Biological Wastes

Soil gets large quantities of human, animals and birds excreta, which constitute major sources of land pollution. The pathogens from their excreta are transmitted from animals to soil and then from soil to man. Thus biological agents are highly responsible for heavy contamination of soils and crops by the pathogens. Many pathogens are extremely persistent and are strongly resistant to attacking agents.

Fertilizers

Fertilizers are one of the most important inputs for increasing agricultural production. These fertilizers crowd out essential nutrients present in topsoil layers, where the microbes enrich the humus and initiate plant growth. The fertilizer-enriched soil cannot support microbial flora. Keeping in view its importance, it was declared as essential commodity under Essential Commodities Act, 1955. Fertilizer means any substance used or intended to be used as fertilizer of the soil or crop.

i.   It is reported that there is a decline in protein content of 25% when corn, maize, groundnut, wheat crops were grown on soil fertilized with NPK fertilizers. The subtle balance of amino acids with the protein molecule is also disrupted degrading the carbohydrate and protein quality leading to malnutrition.

ii. Potassium fertilizers in soil decrease the valuable nutrient ascorbic acid and carotene in vegetables and fruits.

iii. Fertilized soil produces bigger sized vegetables and fruits, which are more prone to pest, insects and diseases.

iv. Excessive use of nitrogenous fertilizers in land leads to accumulation of nitrate in the soil, which inturn is transferred to man through plant cause diarrhoea, and cyanosis (blue jaundice) in children.

v.   Generous use of nitrate fertilizers has raised the nitrate levels in soils to dangerous amount, especially to young infants susceptible to fatal disease called methemoglobinemia or blue baby syndrome.

vi. Phosphatic fertilizers like DAP are considered detrimental to crop production.

vii. Cereal crops like jowar, maize and pearl millet grown on alkaline soil absorb higher amount of fluorides and are responsible for fluorosis.

viii. Fertilizers are contaminated with other synthetic organic chemicals thereby polluting the soil water. These fertilizers enter the water system contributing to eutrophications, i.e., excessive growth of algae and aquatic plants producing evil odours. Excess use of fertilizers intensively reduces the ability of plants to fix nitrogen.

Pesticides

Pesticides not only pose a potential hazard to men, animal, fish and livestock, but they severely affect the desired yield of crop and spoil. Even the accepted dose of pesticides creates deleterious effect on soil fertility. Many of the pesticides application may be unnecessary and are economically unsound. A range of alternative methods of pest controls to be used in organic farming is given below.

1. Deep ploughing the fields during summer season helps in killing pests, larvae and eggs.

2. Clean cultivation by destruction of weeds and other alternate hosts breaks the carry over of the pests in succession, which considerably reduces the pest numbers.

3. Adapting crop rotations to avoid carry over of pests from one season to next season.

4. Change in time of sowing.

5. Draining of water out of fields at times of pests growing in number.

6. Use of resistant varieties.

7. Growing of trap crops.

8. Release of parasites and predators.

9. Use of pheromone traps and light traps.

10. Use of biological insecticides.

11. Use of mechanical weed control.

12. Cover cropping to control weed-seed germination.

Organic agriculture is a viable alternative because it enlivens the soil, strengthens the natural resource base and sustains biological production at levels to commensurate their carrying capacity of the managed agro-ecosystem.

Some of the adverse effects of pesticides are the following.

i. They are retained in soil for long duration and affects crops, vegetables, cereals and fruits and taint them to such an extent that they are not usable.

ii. Various crops such as vegetables, fruits, rice, grain, wheat, gram, maize, etc. are known to contain significant amounts of DDT, BHC, and other organo-chlorine pesticides. They persist in the soil producing long-term effects on vegetative cover. PolyChlorinated Biphenyls (PCBs) are the most dangerous having a half-life period of 25 years in soil. They accumulate in the soil and ultimately enter into human bodies. PCBs cause deformities in foetus, nervous disorders, liver and stomach cancer in animals.

iii. Persons who used vegetables contaminated with 0.5 gm or more PCBs developed darkened skin, eye damage and severe acne.

iv. Pesticides like DDT, endrin, dieldrin, heptachlor, etc. are known to seep gradually through soil into groundwater and eventually contaminate public drinking water supplies.

Controlling pesticide pollution in soil   Agricultural pesticides hit the aquatic and terrestrial ecosystem ranging from acute toxicity to invisible deleterious effects. Some of the possible methods to control the pesticides contamination in soil are as detailed below:

i. Less selective and harmless pesticides can be used instead of highly persistent ones.

ii. Non-selective persistent pesticides such as DDT and BHC must be phased out of use.

iii. Chlorinated pesticides should not be used.

iv. Repeated applications of even selective pesticides could be avoided.

v. Pesticides use should be properly regulated.

vi. Industrial effluents containing pesticides can be treated before they are disposed into the soil.

vii. General public, farmers, field workers and agriculturalists should be educated regarding the use of pesticides and their environmental hazards.

viii. A new policy for future is to develop a joint biological and chemical approach to pest's control.

REMEDIAL MEASURES TO SOIL POLLUTION

Some of the remedial measures for the soil pollution are the following.

i. Proper dumping of unwanted materials.

ii. Production of natural fertilizers.

iii. Use of biopesticides.

iv. Proper hygienic conditions.

v. Recycling and reuse of wastes.

vi. Ban on toxic chemicals.

vii. Compulsory reporting of insecticide hazards.

viii. Evaluation of pesticides on adverse effects on soil and crops.

ix.   Establishing laboratories to detect the microquantities of pesticides in soil.

x.   Plantation of trees.

xi.   Treating industrial wastes for the required standard.

xii.   Treating domestic wastewater and reusing for agriculture.

Methods to minimize soil pollution   Some of the methods used to minimize the soil pollution are the following.

i.   Separate garbage bins can be used to collect different varieties of wastes with the coordination of people in separate colour-coded bins.

ii.   Papers should not be mixed with glass or plastic which are difficult to recycle.

iii.   Encouraging the people by ways of subsiding the wastes.

iv.   Tax exemptions are also beneficial to enhance recycling of wastes.

v.   Making use of recycled paper instead of fresh ones.

vi.   By reducing the creation of wastes, recovering, recycling and reusing potential wastes, the amount of water can be reduced effectively.

vii.   Genetic engineering has been developed in recent years. Using this technology, plants could be made to be resist to many diseases and insect pest. New varieties could be developed that could live in salty soils.

ORGANIC FARMING

With an increase in population, compulsion would be not only to mobilize the agricultural production, but also to increase further in a sustainable manner. Therefore, there is a need to study the past trends in inputs usage like pesticides and fertilizers, which are major components in crop production. The organic matter content of soils has come down to less than one per cent because of indiscriminate use of chemical fertilizers for decades. In addition, the use of pesticides leads to pest resurgence and difficult to control weeds species. Hence, the expectation that organic farming by reverting to the use of manures, green manures, urban wastes, rural wastes, etc. can bring sustainability to the agriculture with eco-friendliness.

Organic farming is the production of crops that are free from synthetically compound fertilizers, pesticides, growth regulators and livestock feed additives. To the maximum extent feasible, organic farming systems rely upon crop rotation, crop residues, animal manures, legumes, green manures, and mechanical cultivation to maintain the soil productivity and to supply plant nutrients, and to control insects and weeds.

Types of Organic Farming

Pure organic farming   It includes use of organic manures and bio-pesticides with complete avoidance of inorganic chemicals and pesticides.

Integrated farming   It involves integrated nutrient management and integrated pest management. In this type, local resources are effectively recycled by involving other components such as poultry, fishpond, mushroom, goat rearing, etc. apart from crop components. It is a low input organic farming.

Vermicompost   It involves the rearing of earthworms on waste materials such as cow dung, agricultural residues, kitchen garbages, etc. The earthworm used for vermicomposting is of "epigeic" type (African variety).

Preparation of a vermicompost   The following are the steps involved in the preparation of vermicompost.

   i.   Semi-permanent thatched roofing sheds are formed to protect the worms from birds, rats, etc.

   ii.   All around the inside wall, chambers of size $10'' \times 3'' \times 1\frac{1}{2}''$ are constructed using bricks. The bottom is filled with broken bricks, followed by red soil of about by about $2''$ thick. Over the soil, biodegradable wastes like coir pith, leaf litter, etc. are spread for a thickness of $6''$. Over that $3''$ layer of cow dung is prepared. Figures 15.4 and 15.5 show the vermicomposting yard and the vertical sectional view of the vermicomposting bed respectively.

   iii.   Keeping the moisture content not exceeding 55%.

   iv.   Feed appropriate worms in it.

   v.   Each worm can compost 6 kg of wastes. In 6 months, each worm can produce 400 worms. About 800–1000 worms will be found in 1 kg of wastes. To produce 1 ton of vermicompost, 1500 worms are needed. Within 90–100 days, vermicompost in granular forms will be produced and is ready for use.

Figure 15.5   Vermicomposting bed with shed

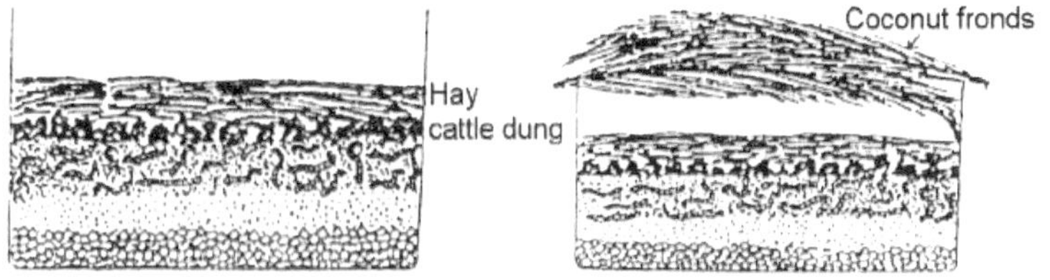

Figure 15.6   Vermicomposting in bins

The advantages of vermicomposting are the following.

  i. Nitrogen-fixing bacteria are present.

  ii. It can dissolve phosphate.

  iii. NPK value is considerably high.

  iv. Chemical fertilizers can be reduced gradually, and within 3 years a complete organic forming can be maintained.

  v. It has the property to retain moisture in soil and hence water-feeding interval can be increased and in other terms, water usage will be minimal.

  vi. It has good iron and other mineral contents thereby increasing the soil property.

   vii.   It contains enzymes, hormones, and vitamins, which are essential for better growth of plants.

   viii.   Air circulation and water penetration are made easy because it loosens the soil, which allow the roots to go deeper and deeper easily.

   ix.   It can control hookworms. As a result, soil fertility and its environment with microorganisms are properly kept.

ALTERNATIVES FOR CHEMICAL FERTILIZERS

In India, the use of organic manure in subsistence forming is an age-old practice. For substituting the chemical fertilizers various forms of organic manures and biofertilizers are explained below.

Farm Yard Manure (FYM)

Cow dung is an important source of plant nutrients. FYM is composed of dung, urine, bedding and straw. FYM contains approximately 5 to 6 kg N, 1.5–2 kg phosphorus and 5–6 kg potash /ton. It builds up the soil health considerably.

Green Manure

It is considered as a good source of "N" and it increases the availability of P, K secondary and trace elements to the soil.

Coir Pith

The annual production of coir pith in India is about 7.5 million tons. Preferably biodegraded and amended coir pith can serve as a substitute for FYM or similar organic manure, *Pleurotus sojorcaju*, *Aspergillus* and *Trichoderma* are found to be potent in degrading the coir pith.

Vermicompost is 5 times richer in N, 7 times in P, 11 times in K, 2 times in Mg, 2 times in Ca and 7 times in antimony than ordinary soil. It is a rich source of vitamins and growth hormones like gibberillic acid, which regulates the growth of plant and microbes. The compost prepared by using earthworm is called vermicompost.

Biofertilizers

The living cells of different types of microorganisms, which have an ability to mobilize nutritionally important elements from non-usable to usable form are called biofertilizers, e.g. *Rhizobium*, *Azotobacter*, *Azospirillum*, blue-green algae, *Azolla*, *Mycorrhizae*, and phosphate-solubilizing bacteria. These microorganisms require organic matter for their growth and activity in the soil and provide valuable nutrients to the plants in the soil. They influence the availability of major nutrients like nitrogen, phosphorus, potassium and sulphur to the plants.

SOIL EROSION

Human activities such as modern agriculture practices, replacing natural forest and grassland covers by construction works, desertification and over grazing, etc. have accelerated the natural soil erosion. Due to soil erosion, there is acute siltation of water system chocking estuaries and harbours. Top fertile soil is lost due to erosion. Eroded soil gets deposited on riverbeds increasing their water land leading to devastating floods. Waterlogging have resulted in loss.

Control Measures

The actual art of soil conservation is with heaping the soil intact and maintaining the soil nutrients at certain desired level. The steps involved are the following.

1. Increase in soil resistance
2. Land should be fairly levelled
3. Crop rotation
4. Strip cropping
5. Contour farming
6. Constructing dams, drains and widening of gullies can channel excess run off
7. Afforestation
8. Plastic filter fabrics.

Marine pollution is advancing as a major problem on earth. But there are billions of people who do not believe that ocean problem is a problem. Most of us do not even know what types of pollutants reach the oceans and the effects that pollution has on the oceans but we continue to dispose of chemicals, sewage and garbage into it. In this chapter, we will look at the various ocean pollutants and the potential impacts that they have on the ocean animals.

TOXIC OCEAN POLLUTANTS

Toxic pollutants in an ocean ecosystem have massive impacts on the plants and animals. Some of the major toxic pollutants of ocean are the heavy metals, dioxins, polychlorinated biphenyls (PCBs), polyaromatic hydrocarbons (PAHs), etc.

Heavy metals such as lead and mercury from industrial effluents get collected in the tissues of top predators such as whales and sharks, and can cause birth defects and nervous system damage.

Dioxins from the pulp and paper bleaching process can cause genetic chromosomal problems in marine animals and may even cause cancer in humans. PCBs (polychlorinated biphenyls) typically cause reproduction problems in most marine organisms. PCBs usually appear from old electrical equipment.

Polyaromatic hydrocarbons (PAHs) are another source of marine toxic pollution and typically come from oil spillage and burning wood and coal. These PAHs are responsible for causing genetic chromosomal aberrations in many marine animals.

Finally, low-level radiation poisoning is also possible in the ocean environment. Scientists know very little about how radiation affects marine organisms. Some marine species such as a population of Beluga whales living in the Saint Lawrence River area in Eastern Canada are in serious trouble because of marine toxic pollution. These Beluga whales are the victims of ocean pollution ranging from PCBs to heavy metals as well as other pollutants. However, toxic pollution is only the tip of the iceberg in terms of total ocean pollution.

MARINE GARBAGE

Marine garbage disposal is another major form of ocean pollution. The world's oceans are a virtual dumping ground for trash. Sometimes the garbage include junked out fishing nets, plastics, general household garbage and even scraps like light bulbs. In one case, an island 300 miles from the nearest inhabited island (and more than 3000 miles from the nearest continent) had 950 pieces of garbage ranging from plastics to tin cans. Garbage in the oceans is a serious issue as fish entangle themselves in fishing nets and animals sometimes eat trash products and die.

Each year, there are numerous examples of dolphins, sharks and whales entangling themselves in fishing nets and dying from oxygen starvation. It is possible to clean garbage from the oceans if humanity quits using it as a garbage dump.

The following pictures will give some indication about potential garbage pollution in the world's oceans (Figure 16.1).

Figure 16.1   Waste plastics in ocean bottom

Marine garbage can often be ingested into animal stomachs, plastic pop tab rings that accidentally strangle animals, and hence controlling this form of pollution is important to maintaining a healthy ocean ecosystem. Even simple plastic bags can have large pollutive impacts within the ocean. For example, a deceased sperm whale was found to have a party balloon blocking its digestive system. Not able to process its food, it starved to death. Plastics can also have negative impacts to boats if they accidentally plug the water intake lines. There are instances known where plastic bags get plugged into water intake lines used for cooling the boat motors. This resulted in overheating of the boat motor, causing some serious engine damage.

OCEANIC SEWAGE

Sewage is yet another major form of ocean pollution. Typically the problem with sewage is that it causes massive nutrient loading in the ocean ecosystem. When nutrient loading occurs there will often be algal blooms in the water leading to the loss of dissolved oxygen. Finally, the oxygen depletion causes death of many organisms.

Other problems associated with sewage include parasites/bacteria that require the closing of coastal beaches and poisoned shellfish fisheries. For the most part, the cities in the developed world have sewage treatment facilities but many of the cities in poorer areas have little to no sewage treatment.

The sewage pollution increases with an increase in the population. The following pictures illustrate some of what can reach the ocean through our sewage (Figure 16.2).

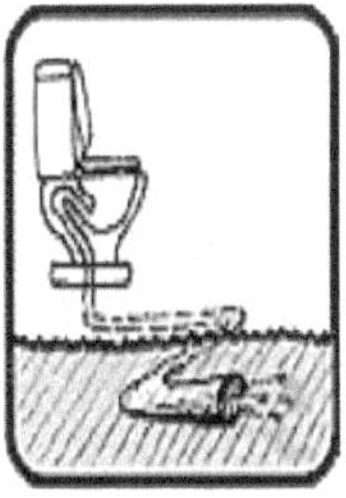

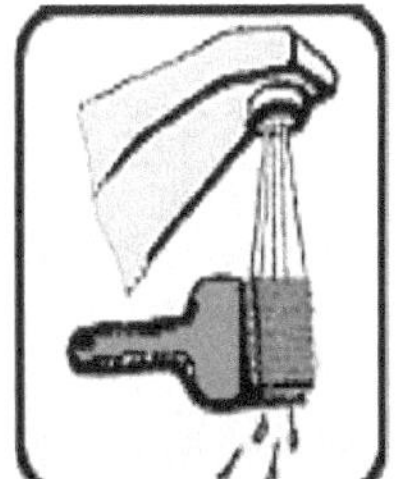

| Sewage effluent | Pollutant entering oceanic floor | Washed-down paints |

Figure 16.2   Wastewater entry into ocean

The kitchen wastes and domestic sewage industrial effluents and even chemicals such as paints and fertilizers that we dispose of down the drains eventually reach the ocean. Eventually all of this sewage pollution adds up to the serious problems of lack of available oxygen for organisms leading to death of the life in ocean. The sources and effects of marine pollutants are listed in Table 16.1.

NON-POINT SOURCES

The other non-point sources of pollution from various forms of run-off, e.g. run-off from farming (fertilizers manure), industrial run-off (heavy metals, phosphorous), urban run-off (oils, salts, various chemicals) and atmospheric fallout of airborne pollution. The non-point pollution is the hardest to control because it comes from so many sources. Hence, it is clear that the non-point pollution is pollution that comes from a variety of sources such as road run-off all the way to agricultural run-off in contrast to the point pollution where comes from a direct source like a factory outfall pipe.

NOISE POLLUTION

Noise pollution in the oceanic surface also leads to several problems by causing stress to the marine flora and fauna.

SOLUTIONS

◇ Humans can make an impact on preserving the oceans, provided they make an effort to lessen the pollution.

◇ Industrial pollution is not as bad as in the developed countries, as we have new techniques and better waste treatment than before.

◇ New laws and regulations make it difficult for people to dump their trash into the oceans though inevitably some dumping will always occur. One idea is to have beach cleaning days where everyone chips in to clean the trash off the beaches. By cleaning up the trash on beaches we lessen the chances of accidentally killing animals and it makes the beaches look better.

◇ Reduction of sewage is possible through the installation of better sewage treatment facilities for the world's cities. Developed countries like Canada and the United States as well as Western Europe should

Table 16.1 Sources and effects of marine pollution

Types	Primary source/cause	Effect
Nutrients	Run-off approximately 50% sewage, 50% from forestry, farming, and other land-use. Also airborne nitrogen oxides from power plants, cars, etc.	Feed algal blooms in coastal waters. Decomposing algae depletes water of oxygen, killing other marine life. Can spur algal blooms (red tides), releasing toxins that can kill fish and poison to people.
Sediments	Erosion from mining, forestry, farming and other land-use; coastal dredging and mining.	Cloud the water; impede photosynthesis below surface waters. Clog the gills of fish. Smother and bury the coastal ecosystems. Carry toxins and excess nutrients.
Pathogens	Sewage, livestock.	Contaminate coastal swimming areas and seafood, spreading cholera, typhoid and other diseases.
Alien species	Several thousand per day transported in ballast water; also spread through canals linking bodies of water and fishery enhancement projects.	Out compete native species and reduce biological diversity. Introduce new marine diseases. Associated with increased incidence of red tides and other algal blooms. Problem in major ports.

(Contd.)

Table 16.1 (Continued)

Type	Primary source/cause	Effect
Persistent toxins (PCBs, heavy metals, DDT, etc.)	Industrial discharge; wastewater discharge from cities; pesticides from farms, forests, home use, etc. seepage from landfills.	Poison or toxinscause disease in coastal marine life, especially near major cities or industry. Contaminate seafood. Fat-soluble toxins that bio-accumulate in predators can cause disease and reproductive failure.
Oil	46% from cars, heavy machinery, industry, other land-based sources; 32% from oil tanker operations and other shipping; 13% from accidents at sea; also offshore oil drilling and natural seepage.	Low level contamination can kill larvae and cause disease in marine life. Oil slicks kill marine life, especially in coastal habitats. Tar balls from coagulated oil litter beaches and coastal habitat.
Plastics	Fishing nets; cargo and cruise ships; beach litter; wastes from plastic industry and landfills.	Discard fishing gear continues to catch fish. Other plastic debris entangles marine life or is mistaken for food. Plastics litter beaches and coasts and may persist for 200 to 400 years.
Radioactive substances	Discarded nuclear submarine and military wastes; atmospheric fallouts; also industrial wastes.	Hot spots of radioactivity. Can enter food chain and cause disease in marine life. Concentrate in top predators and shellfish, which are eaten by people.
Thermal	Cooling water from power plants and industrial sites.	Kill off the corals and other temperature-sensitive sedentary species. Displace other marine life.
Noise	Supertankers, other large vessels and machinery.	Can be heard thousands of kilometers away under water. May cause stress and disrupt marine life.

assist the poorer countries in installing sewage treatment facilities. Reducing harmful sewage discharges would be a major start in helping to clean the oceans of pollution.

◇   Many areas of the world have reduced non-point pollution through proper recycling of used oil and paint products. (In the past, people simply dumped used oils and paints into the sewer system where they would do serious damage to the water.) Pollution will still occur but with effort and determination it is possible to reduce its impact on the oceans.

REVIEW QUESTIONS

1. Define air pollution. Name the sources of air pollution. How are air pollutants classified?
2. What are the effects of air pollution on economics?
3. Describe the effects of air pollution on plants and animals.
4. Enumerate the approaches to air pollution control.
5. What are the control techniques for particulate contamination?
6. Discuss various treatment processes for the control of gaseous contaminants and explain any one.
7. Explain the working of fixed-bed absorbers for removal of gaseous pollutants.
8. Discuss the effects and sources of CO_2.
9. Describe the characteristics, sources and effects of CO_2.
10. Describe the characteristics, sources and effects of sulphur dioxide.
11. Discuss ozone as a pollutant and non-pollutant.
12. What are the primary sources of nitrogen oxides and their ill effects?
13. What are water pollutants? Discuss the sources.
14. Suggest various control and remedial measures to remove water pollution.
15. What are the characteristics of wastewater and how do these characteristics change the environment?
16. What are the various methods for treatment of wastewater?
17. What are the unit processes involved in wastewater treatment? Discuss any one method of treatment of wastewater.
18. What are the sources of soil pollution and what are its ill effects?
19. Enumerate the ill effects of fertilizers and pesticides to soil.
20. Describe the measures to control soil pollution.
21. Is organic farming a sustainable way of bringing the soil to future generations? Comment on this.
22. Enumerate the measures to control soil erosion.

23. What are the alternatives for chemical fertilizers?
24. What are the sources of noise in the atmosphere? Enumerate the ill effects of noise on humans.
25. What are the legal aspects in controlling noise?
26. Discuss the noise-control measures.
27. What are the sources of thermal pollution?
28. Discuss the effects of thermal pollution on the environment.
29. What are the types of radioactive waste and how are they harmful to the human environment?
30. Describe the methods of disposing nuclear wastes.
31. What are the characteristics and sources of solid waste?
32. Classify solid waste according to use and describe their sources.
33. What are the effects of solid waste on the community, if not managed properly?
34. Describe the activities of solid waste management.
35. What are the engineered techniques in management of solid waste?
36. Describe different disposal methods of solid waste.
37. How are industrial solid wastes handled?
38. What are manmade disasters and natural disasters? Describe different phases of emergency management in the case of disasters.
39. What is flood? What precautions are taken during and after a flood?
40. Define earthquake. What are causes and effect of earthquake?
41. How are the zones of earthquake important in development activities?
42. Comment on the seismic zones of India.
43. What are cyclones? What is cyclone watch?
44. Describe the causes for different types of cyclones.
45. How are landslides formed?
46. What are the likely impacts of landslides and the contributing factors to vulnerability?
47. What are preparedness measures and mitigation measures to landslides?
48. Describe the post-disaster measures to be adopted for landslides.
49. How can an individual contribute to the prevention of pollution?

The present-day technological advancements and increase in per capita income are still considered as the symptoms of progress and development, all at the expense of ruthless exploitation of nature. Though the exploitation of natural resources is inevitable for the development and progress of mankind, we now need a rethinking. Earth has limitations and natural resources are also limited. What is taken from the nature should be returned for its survival.

World organizations have realized that economic development cannot succeed unless sufficient attention is given to the natural environment. During development planning, only those technological developments with minimum environmental hazards should be adopted in order to sustain the environment for future generations.

ENVIRONMENTAL ISSUES

The ultimate aim of development is to make all organisms including human beings happier. But our present patterns of progress and development have resulted in human miseries and sufferings resulting in the denial of control of people over their lives. Their basic rights to drink pure water, breathe unpolluted air, eat unadulterated food and enjoy natural resources are denied.

The following are the social issues of the present world.

- ◇ Depletion of non-renewable resources like fuels and minerals.

- ◇ Qualitative degradation of renewable resources like plant resources, animal resources, marine resources, water, air, soil, forests, etc.

    ◊   Emergence of diseases like obesity, cardiovascular ailments, cancer, alcoholism, drug addiction, and so on.

    ◊   Overcrowding and overpopulation in cities.

    ◊   Unemployment for the teeming millions.

If development destroys our culture and brings massive human tragedy as in Bhopal and Chernobyl, it is not the development. Two-thirds of humanity have become victims of the present pattern of development. Women and children are the worst sufferers. Development of poverty rather than happiness is the outcome. Most people are hungry today than ever before, and their number is increasing.

"We are stealing the future from our children. This generation is passing heavy ecological, social and economic costs to its successors."

Today man is so alienated from nature that 80 per cent of our life is spent in an artificial environment. This human–nature relationship simply mirrors the degrading human–human relationship in the society.

The possible solutions for the above-discussed environmental issues are detailed in the following sections.

RESOLVING ENVIRONMENTAL ISSUES

Sustainable Development

Sustainable development is the process of social and economic betterment that satisfies the needs and values of all interest groups, while maintaining the natural resources.

"Meeting the needs of the present without compromising the ability of future generations to meet their own needs" is called the sustainable development. Sustainable development, like environmental management, is not easily defined. It is the application of science, technology and environmental knowledge to world development. Some definitions of sustainable development are the following.

1.   It is the development based on the principle of inter-generational, inter-species and intergroup equity.

2.   It integrates environmental care management and development.

3. It is an environmental "handrail" to guide development.

4. It is a change in consumption patterns towards more benign products, and a shift in investment patterns towards augmenting environmental capital.

5. It is a utilitarian conservation adopted with a long-term view along with an internationalist stance.

6. It improves the quality of human life while living within the carrying capacity of supporting ecosystems.

7. It is a process that seeks to manifest a higher standard of living (however interpreted) for human beings who recognize this cannot be achieved at the expense of environmental integrity.

Sustainable development is closely linked to the carrying capacity of an ecosystem that determines the limits to economic development. Carrying capacity of a specific ecosystem is the maximum rate of resource consumption that can be sustained indefinitely in that specific area, and over-exploitation of natural resources above this maximum will lead to depletion and ecological degradation.

Carrying-capacity-based planning ensures sustainable development. The carrying-capacity-based planning process, innovative technologies for enhanced material and energy affectivity of production and consumption, structural economic changes towards less resource-intensive sectors, and preventive environmental management through increasingly interventionist policies are some of the strategies for reconciling developmental goals with ecological capabilities.

One problem faced by our environmental managers is that the goal of sustainable development is not fully formed and its fundamental concepts are still debated. Figure 17.1 lists some of the principles for shifting to more environmentally sustainable economy.

The concept of sustainable development, although had appeared in the 1970s, was widely disseminated in the early 1980s by the "World Conservations Strategy" (IUCN, UNEP and WWF, 1980), which called for the maintenance of essential ecological processes, preservation of biodiversity, and sustainable use of the species and ecosystem.

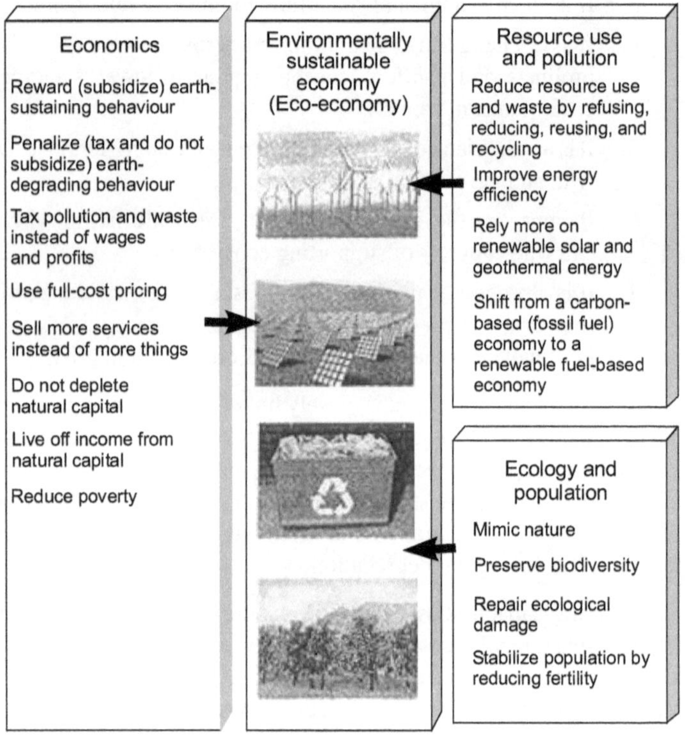

Figure 17.1   Principles for shifting to more environmentally sustainable economy

Thus, sustainable development is a moderate position between the extremes of no growth versus unlimited growth, and is based on the use of renewable resources in harmony with ecological systems (Figure 17.2).

Sustainability means that there is no long-term environmental cost to nature in the manufacture of a product or the provision of a service because the base materials from which such products and services are made and provided can be replaced in a reasonable period of time by natural processes.

The increasing concern with environmental problems in developing countries and the failure to relate these problems to development issues led to the establishment of the United Nations Commission on Environment and Development in November 1983. This commission, under the

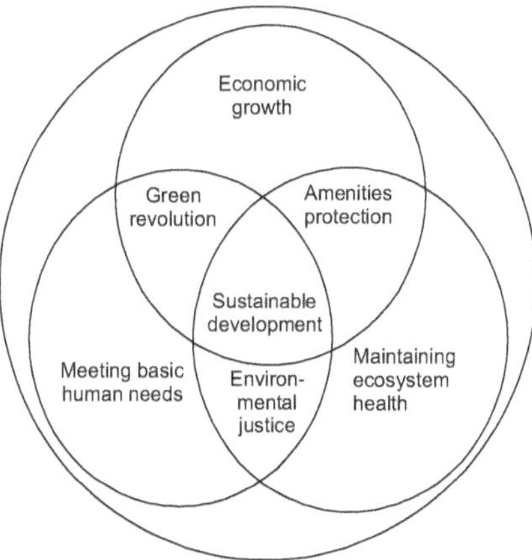

Figure 17.2   Integration of human needs, ecosystem, health and sustainable economic growth

energetic leadership of Norway's Prime Minister Mrs. Brutland, consisted of twenty-two people from both the developed and developing countries. The objective of the commission, according to its interim statements, was to focus on the causes of environmental problems rather than the effects of environmental degradation. The commission concluded that the very nature of economic development must change the poverty, and the cumulative negative impacts of human activities are to be reduced dramatically. In other words, the world must have sustainable development.

Goals of sustainable development   The goals of the sustainable products are the following.

- ◇   The survival and well-being of human beings and all other species.

- ◇   The survival and well-being of the earth, natural resources and all factors that support life on earth.

- ◇   Provide a high-quality life for present and future generations.

- ◇   Achieving the above without exceeding the environment's ability to recycle wastes, provide resources and support a rich diversity of life.

Human beings are the most powerful species on earth and hence they can save and enhance life on earth. Life of all other organisms and the quality of the nature are related to their existence and therefore, today it is a greater urgency and responsibility on their part to maintain all other species and nature. Because of these goals, sustainable development is not always compatible, progress and development, knowingly or unknowingly, cause extinction of other species, depletion of resources and damage to the environment.

This requires planned utilization of resources and protection of natural wealth. Human needs should be met both by increasing production potential through sustainable development and by ensuring greater economic opportunity for all. This also requires reduction and removal of economic inequality and inefficiency, as both contribute to ecological unsustainability. For example, subsidies have encouraged inefficient use of pesticides and fertilizers.

Certain values should be promoted to encourage sustainable rates and types of resource consumption and waste disposal. Some changes in our attitudes, life styles and accepting behavioural patterns are required based on environmental ethics. The habitats of other species should be maintained and the sustainability in use of other species must also be ensured.

We need extreme love, care and consideration for all other living beings. People should have their own way of choosing to meet the above conditions. These conditions are however not met with strategic directions and targets for sustainable development. To achieve the needed goals the following are to be considered.

- ◇ Stop imitating the high-energy-consuming and resource-depleting western pattern of development.

- ◇ Stop developmental patterns based on loans, subsidies, imported resources and borrowed techniques.

- ◇ Move towards an eco-development using indigenous materials suited to our culture and traditions at low costs.

- ◇ Reduce annual emissions of CO_2, SO_2, NO_2.

- ◇ Shift from fossil fuel to solar, wind, mini- and micro-hydel, and biogas.

- ◇ Reduce energy consumption.

- ◇ Reduce generation of wastes.
- ◇ Use public transport over private transports.
- ◇ Increase energy efficiency of all transports.
- ◇ Try an economic development with conservation of ecological processes and biological diversity.
- ◇ Reduce the rate of tropical deforestation.
- ◇ Reduce species extinction.
- ◇ Retain as much natural environment as possible.
- ◇ Act for basic rights of women and children.
- ◇ Conserve the basic rights of the poor, weaker and indigenous sections of the society.
- ◇ Involve people to represent and participate.
- ◇ Plan to reduce population growth.
- ◇ Reduce inequalities based on social justice.
- ◇ Maintain culture and values.

Guidelines that could be followed in our day-to-day life

- ◇ Develop the attitude of using recycling, reusing and refilling products, e.g. bottles.
- ◇ Save energy.
- ◇ Use a travel cup in the car instead of disposable cups.
- ◇ Use fluorescent light instead of colour bulbs.
- ◇ Keep noise levels from radios, television sets or household machines to the minimum.
- ◇ Always reheat thoroughly, cooked foods that have been stored for a long time.
- ◇ Use paper plates instead of plastic.
- ◇ Buy stationery, cards and wrapping papers from recycled paper.
- ◇ Avoid buying "over packed" goods and foods whenever possible.
- ◇ Keep public places such as bus stand and railway station clean by providing with bins and comfort stations.

◇   Buy recyclable bond and computer paper.

◇   Keep students informed and regularly updated on local, regional, national and global environmental issues.

People who have positive ideas about man and nature should get together in small groups to a mass movement for the protection of the mother earth. They should study local environmental problems and suggest solutions. "Think globally and act locally", should be the motto.

Environmental understanding among the masses has always been a powerful force. Whenever and wherever they are organized in a group to protest and challenge, the Governments have been compelled to act. Short-sighted developmental activities have been discarded. The mass movement that led the battle against the Silent Valley, hydroelectric project compelled the Government to drop the project. In Europe, the Governments could not go on with nuclear energy projects against the peoples' will. The ocean parties have played a significant role against the nuclear arms race.

Here are a few examples of some movements.

i.   **Save the Silent Valley Campaign**   This was the first major campaign against a dam in India which started in the early 1980s. It successfully saved a genetically rich and one of the last remaining rainforests in Kerala from being submerged. The campaign was spearheaded by the Kerala Sastra Sahitya Parishad.

ii. **Save the Taj Campaign**   Environmentalists feared that pollution from the Mathura Refinery, located 40 km away, could damage the Taj Mahal. The authorities took precautionary measures to protect the monument from any sign of deterioration.

iii. **Save the Western Ghat March**   This padyatra, jointly organized by a number of environmental groups in 1988, covered over 1300 km across the state of Maharashtra, Goa, Karnataka, Tamil Nadu and Kerala. The march focused its attention to the environmental problems of the Western Ghats.

iv. **Save Water, Save Life**   This was organized by the national fishermen's forum. This padyatra spanned the entire coastline of India from West Bengal in the east and Gujarat in the west to Kanyakumari in the south. It drew attention to the problems of water pollution and the over-exploitation of fisheries by commercial trawler—the two major problems faced by the country's traditional fisher folk.

Rio Declaration for sustainable development   The United Nations Conference on Environment and Development (UNCED)—Earth Summit—was organized in Rio de Janeiro from 3 to 14 June, 1992. The summit brought together more than 100 heads of state and ministers or other top officials from 172 governments, as well as several thousand members of non-governmental organizations (NGOs) and more than 8000 media representatives. It made history particularly for people in developing countries. For the first time, a consensus was reached at the highest level of governments to adopt a radically new people-centred approach to fighting poverty and raising living standard. Approach that creates a global partnership for sustainable development by bringing together governments inter-governmental organizations, NGOs, public and private institutions, and interested citizens.

The consensus consecrated recognition by the heads of state that poverty alleviation must go hand in hand with environment protection and sustainable development. The carrying capacity of the earth and the carrying capacity of its people were intertwined to spearhead Agenda 21: the most far-reaching practial measures for human development ever agreed upon by governments. This contains 115 targeted programme areas and clearly specifies priority action, driving forces and the essential means at all levels of endeavour—local, national, regional and international.

Agenda 21 is underpinned by the "Rio Declaration"—a set of principles outlining the rights and responsibilities of states in achieving sustainable development.

Standard measures in environmental rules should, as far as possible, be based on consensus.

URBAN PROBLEMS RELATED TO ENERGY

Energy is the primary input for almost all economic activities and is therefore vital for improvement in quality of life. Its use in sectors such as industry, commerce, transport, telecommunication, wide range of agriculture and household services has compelled us to focus our attention to ensure its continuous supply to meet our ever-increasing demands.

Coal, oil, gas and water constitute the main sources of energy in our country. Energy is obtained mostly from coal (56%) and petroleum (32%) and other sources are natural gas and water. Apart from this energy, a large amount of traditional energy sources in the form of fuels, wood, agriculture and animal residues are used.

Energy Demand

Energy demand is not an exception to the economic theory of limited means and unlimited wants. The pace of exploitation of the energy resources has been growing over time and has resulted in gradual depletion of scarce reserves. Today, energy has become a key factor in deciding the product cost at the microlevel as well as indicating the inflation and the debt burden at the macrolevel. Energy cost is a significant factor in economic activity, at par with factors of production like capital, land and labour. The imperatives of an energy shortage situation call for the energy conservation measure, which essentially means using less energy for the same level of activity. While on one hand the demand for energy is increasing, on the other hand the energy resources are becoming scarce and costlier (Table 17.1). With the increase in the use of energy, the demand supply economics in the industrial scenario and the energy costs escalated year after year.

Table 17.1   The average yearly increase in cost

Year	Average yearly increase in cost (%)
1960s	2–3
1970s	5–6
1980s	8–10
1990s	11–13
2000s	14–17

It is hence said that the energy price doubles once in five years and this naturally leads to the profit-making manufacturing units to turn into a loss-making proposition and the industries were forced to identify the means to conserve to preserve the energy and to stretch the use of the existing source of availability.

Steady increase in gap has compelled the technocrats and decision-makers in the industry to not only develop new measures of energy conservation but also to have a systematic approach towards present trend of energy consumption through energy auditing and application of modern techniques and methods for minimizing energy wastage.

Energy conservation is considered as the quick and economical way to solve the problem of power shortage and also as a means of conserving the country's finite sources of energy. Energy conservation measures are cost-effective, require relatively small investments and have short gestation as well as pay-back periods. The studies conducted by Energy Management Centre, New Delhi, have indicated that there is about 25% potential of energy conservation in the industrial sector.

The country's power demand supply deficit position for different periods is shown in Table 17.2.

Table 17.2 Capacity additions and deficit during various plan periods

Plan	Period	Targeted addition (MW)	Actual addition (MW)	Deficit (%)
I	1951–56	1300	1100	15.4
II	1956–61	3500	2500	35.7
III	1961–66	7040	4520	35.8
IV	1969–74	9264	4579	50.5
V	1974–79	12499	10202	18.4
VI	1980–85	19666	14226	27.7
VII	1985–90	22245	21401	38.0
VIII	1992–97	30538	18000	41.1

There is already a 7% energy shortage of electric power and 17.9% peak demand shortage in the power sector alone. Now India needs two times more of coal, 2–3 times more of oil and 4 times more of electricity. There is 55% gap between indigenous production of crude and demand for petroleum products.

India cannot afford the huge capital costs needed for installation of new power plants, development of coal, mines, strengthening of railway networks, etc. Nor can it afford the increasing drain on precious foreign exchange for oil imports.

The National Energy Conservation Movement has been launched to make consumers aware of their need to conserve energy. All it takes is simple rules and recommended energy-saving procedures in homes, offices, factories and fields where one just does not save energy but also saves money with reduced electricity bills.

Let not the energy crisis change the very way we live tomorrow. Let's take a positive decision on energy conservation and save energy.

The main issues solving the energy problem in urban areas are

◇   Sustainable use of energy

◇   Effective use of energy

◇   Equitable use of energy

Sustainable use of energy   Sustainable energy development would have these following features.

◇   It would provide reliable source of energy.

◇   It would not cause havoc to our global, regional or local environment.

◇   It would maintain quality of environment and fair share of earth's resource.

Effective use of energy   Unless we act and stop wastage of energy and use energy efficiency of which add up to saving of 30%, the country may face a major energy crisis.

Let us contribute a little by saving premium drops of oil, valuable units of electricity. Else, the nation will have to contend with lower rates of growth, falling standards of living, rationing and blackouts. A situation none of us would like to face.

Equitable use of energy   Energy supplies and demands are very difficult to predict because technical, economical, political and social factors are changing. There are also large annual and regional variations in energy management.

All the above factors give rise to the concept of integrated energy management by development and use of alternate energy forms and their equitable distribution for all human land. Urban planning and people participation technologies could not meet the growing power needs.

- ◇ Urban planning networking like mass bus transit, trains and trolleys can provide clean, efficient transportation.

- ◇ If people should ride, they could reduce fuel consumption by 50% or more overnight.

- ◇ If people reduce the number of short trips or ride bicycles, large quantities of fuel would be saved.

- ◇ In general, social solutions offer the quickest and the most cost-efficient methods of conservation. For example, no capital outlays or technological advances are required to turn-off unused lights drive to work in car pools or turn the thermostat down and put on a sweater.

- ◇ Urban planning by formation of policies that are conducive to private investment in green energy generation like alternate energy source like solar cells, wind power, atomic power, ocean power, and tidal and biomass.

- ◇ To chalk out procedure for the transfer, transmission and wheeling of electricity from private grid to the government-managed utility grids.

- ◇ To chalk out policies for accounting and auditing energy use.

WATERSHED MANAGEMENT

Water is necessary for all life. From rainfall to moist soil, to springs, to streams and rivers and to the ocean, water is a link connecting all ecosystems. But many streams that used to have a generally modest flow of clear water throughout the year have changed in character to become torrents of muddy water when it rains and dry channels when it does not. Floods are frequent. Rivers, reservoirs and harbours are filling with sediment. These changes relate to mankind's impact upon the soil and the water cycle.

Natural ecosystems continually purify and recycle water as well as regulate its flow. Humans, however, tend to carry on activities with little understanding of or regard for the natural water cycle. Continuing in this way would threaten our livelihood and perhaps our existence. The failure to add the price of environmental damage to the cost of products generates a market force that increases pollution.

The term watershed strictly refers to the dividing line that separates one drainage basin from another. In India, the term watershed is defined as a land area from which water drains to a given point. Watershed is considered to be synonymous with catchments and drainage basin. Water from a few hectares may drain into a small stream. These few hectares will then be its watershed. This small stream and others like it run into a larger one. The land area drained by the small streams makes up the watershed of the larger stream into which they flow. Watersheds of the small streams are thus the sub-watersheds of the watershed of the larger stream. A watershed may be nearly flat or may include hills or mountains. The people and animals are part of the watershed community. All depend on the watershed and in turn influence the things taking place there both for good and bad. The things that affect the small watershed also affect the larger watershed.

Factors Affecting Watershed

A watershed affects the people in every sphere of life. The sustained productivity of food, fuel, forage, fibre, fruit and water by the management of vital resources of water, soil, vegetation and phenomena like floods and droughts are determined by the nature of watershed functioning. Watershed conditions affect and influence water quality, quantity and regime. These three water parameters are vital for (i) human and animal consumption, (ii) irrigation, (iii) industry, (iv) power generation, (v) recreation, (vi) transportation and many other functions. By manipulating different watershed characteristics, the quality, quantity and regime can be regulated according to requirement.

Erosion and sedimentation   The rate and quantity of sediment produced and transported are of vital importance. Accelerated erosion

    i.   affects adversely the productivity of the land whether it be under agriculture, horticulture, forests or grasses,

   ii.   results in loss of land as a natural resource,

  iii.   results in higher sediment production and delivery into the reservoirs which causes reduction in the life and benefits of the reservoirs,

  iv.   results in siltation of channels, which causes overflows (floods) and damage to land,

   v.   results in damage to machinery (turbines, generators and other plants).

Vegetative production   All lands are dependent upon the availability of the right quality and quantity of water at the right time, which in turn, is dependent upon the conditions of the watershed. The production is also dependent upon the erosion rate and the deposition that takes place in the watershed.

Floods   This again is the function of the watershed and affects the people greatly. Probably, this one single item engages the attention of the people more than any other item. While floods cannot be altogether eliminated, they can be greatly mitigated by sound watershed treatment and management.

Droughts   These could also be mitigated by sound watershed management programmes aiming at moisture conservation and evolving sound land-use programmes. Each individual watershed has a number of distinct characteristics, which affect its functioning with respect to receiving and disposal of water.

Watershed Deterioration

Watershed deterioration takes place due to uncontrolled, unplanned, unscientific land use and activities of humans. These activities are discussed below in detail.

Agricultural land   Cultivation on sloping land without adequate precautions; cultivation along nala or stream banks; cultivation of erosion-permitting crops; over-cropping areas without replenishing soil fertility.

Forest land   Clear-felling on steep slopes; drastic thinning of plantations along slopes; faulty logging, and disturbance of forest floors during removal of the felled trees.

Grasslands   Excessive grazing resulting in disappearance of protective cover, development of cattle tracts into channels, gullies, etc., compaction of soil resulting in lower infiltration rates, etc.

Fire   Intentional or accidental fire results in loss of vegetation, organic matter and microorganisms. Fire disturbs the hydrological behaviour of the watershed very seriously and for a long period.

Shifting cultivation   Shifting cultivation destroys protective and productive vegetation in preference for a very brief period of immediate crop production and results in soil loss and other consequential damages. This practice can cause widespread destruction in the watershed.

Unscientific mining and quarrying   This practice results in exposure and digging up of slopes, which causes considerable damage to the landscape by destroying vegetation. Haphazard disposal of mine spoils blocks drainage channels and covers up pastures or agricultural fields.

Bad road alignment and construction   This can and is causing a lot of dislocation of life in watersheds located in hilly areas. This, precisely, is a major problem of Himalayan watersheds where many new roads (length about 10,000 km) have been made. These contribute largely to coarse sediment to the drainage channels causing blockage of flow.

Non-cooperation of the people   This is the most important factor. Without people's cooperation, no watershed can remain undisturbed, and it is the people's non-cooperation that leads to many problems.

Consequences of Watershed Deterioration

◇ Low productivity of land with respect to food, fuel, forage, fibre and fruits.

◇ Erosion and denudation within and outside the watershed.

◇ Quick siltation of reservoirs, lakes, etc.

◇ Poor-quality water yield due to heavy sediment.

◇ Frequent floods and droughts.

◇ Poor health of people and cattle.

Watershed Management

Watershed management is defined as the rational utilization of land and water resources for optimum production with a minimum hazard to natural resources. It essentially relates to soil and water conservation in the watershed, which means proper land use. Objectives are formulated with the emphasis on the very important aspect of people's participation.

Physical problems   These are the problems that can be easily detected or identified. Steep slopes, bad lands, slide-prone soils, weak geologic formations, etc. can be easily found by observation or with the assistance of existing maps. Problems such as heavy and intense rainfall, excessive run-off and strong winds should be identified from weather and hydrological data or by gathering information and evidence locally.

Resource use problems   Problems such as shifting cultivation, forest destruction, fire, overgrazing, poor road construction and maintenance and uncontrolled mining should be identified and, if possible, the causes should be determined. Clear identification of these problems at the preliminary stage will benefit the follow-up surveys and planning as well as the formation of a realistic policy in the future.

Key problems   The final effects of watershed degradation such as soil erosion, landslides, heavy sedimentation, water pollution, floods, and droughts, etc. must be identified as quickly as possible. This can be done partly by observation and spot-checking, and partly from the data obtained from water resource agencies and local inhabitants. By reviewing or analysing existing information, the history, frequency and extent of these problems can also be determined.

Socio-economic and other problems   Serious socio-economic problems can be major obstacles in carrying out watershed work in developing countries. The serious problems should be identified at the beginning of the planning stage.

In the Fifth Five Year Plan, the watershed approach got wider acceptance and included more diversified programmes. Awareness was generated to develop, conserve, and manage land in a wider perspective taking into account the demands from various sectors.

National policy was adopted to use watersheds of various sizes for the development of land and water resources for conservation as well as reproduction schemes.

In a watershed, the aspects of development with regard to the availability of natural resources and also to provide benefits to the population are taken up in phase.

Optimum production structures are incorporated with a view to obtain a sustained level of production from cultivated land, grazing land and forestlands. Water harvesting in a low-rainfall area and also moderation of floods by different measures are included in planning. The present programme of agricultural development, forestry minor irrigation, animal husbandry and other agro-based industries are included in the programme.

The planning and development of watershed would lead to protection of the ecosystem and result in immediate gains for the people. In watershed development programme, afforestation and plantation crops play a very significant role. The tribals and weaker sections of the society who usually live in most of the areas having denuded or degraded land could benefit from the schemes.

Water Conservation

All of the population may not own a land, but every human being uses water. Water is a precious natural resource for sustaining life and environment. Despite its importance to human survival and prosperity, water remains the most poorly managed resource on earth, squandered and polluted by industry, agriculture, sewage treatment plants and many other abuses. Water is treated as if it comes from an inexhaustible supply. Water conservation is the careful use and protection of water resources without affecting its quality and quantity.

In agriculture   Irrigation generally makes inefficient use of water. One of the most innovative techniques in agricultural water conservation is drip irrigation. This reduces the water needed to irrigate crops by 40–60%. But sophisticated irrigation techniques are prohibitively expensive.

In industry   In the United States, 90% of the water is consumed by industries involved in manufacturing chemical products, paper and pulp, petroleum and coal, primary metal and food processing. Strict pollution control laws provide some incentive for industries to conserve water by recycling water.

In municipal drinking water supply   Repair all, leaks in pipes, water mains, toilets, bathtubs and faucets. It has been estimated that 20%–35% of water withdrawn from public supplies is wasted. Measures and incentives to conserve water must be given top priority.

Cities can reduce their usage of water by recycling or reusing the water. Israel probably has the world's most highly developed system of treating and reusing municipal wastewater. This is done out of necessity, because all of its possible freshwater sources have already been tapped.

We may have to purify and recycle the wastewater. These include consumer education, use of water-saving household fixtures, and economic incentives to save water. In addition to recycling and reuse, cities can decrease water consumption by other conservation measures. Increasing the price of water to reflect its true cost promotes water conservation.

◇   Use low-flow toilets and flush the toilet only when really necessary.

◇   Used water from washing machine can be used for watering plants.

◇   Use of water meter will reduce the usage of water.

◇   Replace lush green garden with decorative rock garden.

RAINWATER HARVESTING

With the growth of population, rapid urbanization and industrialization, increased standard of living and life expectancy, improvement in sanitation, etc. the demand for water is increasing day by day. Groundwater is exploited through open wells, tube wells, springs and horizontal galleries for domestic, industrial and agricultural needs. Indiscriminate development of groundwater often results in groundwater depletion, water quality deterioration and seawater intrusion in coastal zones.

The term "Rainwater Harvesting" implies conservation of rainwater. Collecting and storing of rainwater when and where it falls to use during non-monsoon months is called rainwater harvesting (Figure 17.3). This can be done in two ways:

1.   By diverting the rainwater into tanks, ponds, etc. and conserving it as surface water.

2.   By ingesting it into the soil and conserving it as groundwater.

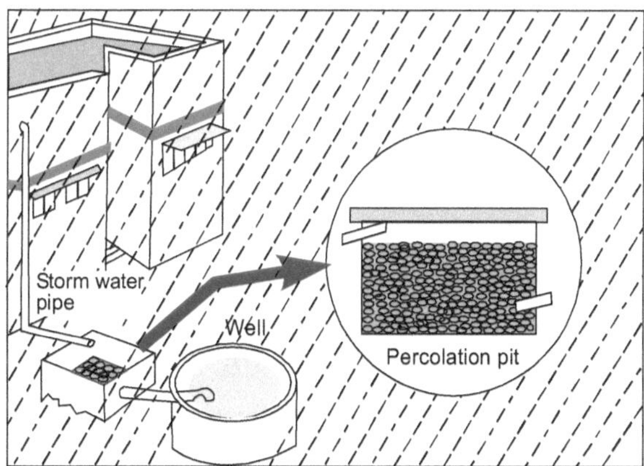

Figure 17.3   Rainwater harvesting from a building

The following are some of the reasons for rainwater harvesting and recharging of groundwater.

⋄   Rainwater is a source of fresh water on earth and if not harvested, run-off gets into the sea and gets wasted without augmenting the surface/subsurface storage.

⋄   Failure to harvest rainwater will flood the low-lying areas and cause a lot of inconvenience.

⋄   It is the solution for water problems where there is inadequate groundwater supply or insignificant surface resources or where rainfall is very intense only over short periods (say 3–4 months in a year).

The rainwater harvesting techniques are broadly classified into two types, each of which is further divided into subtypes as shown in Figure 17.4.

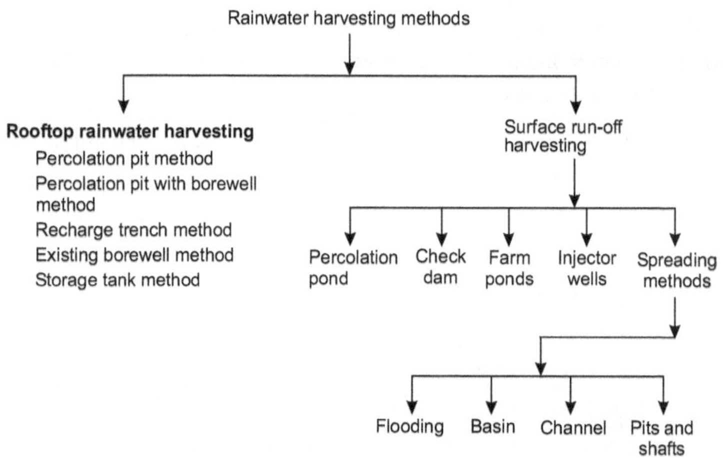

Figure 17.4   Methods of rainwater harvesting

Rooftop Rainwater Harvesting

The rooftops of houses serve as excellent and economical forms of collection centres for rainwater. If properly diverted and used for artificial recharge, it will augment the groundwater table to a sufficient extent.

The initial water at the beginning of the rain has to be used to clean off all the impurities. In the case of a dug well, existing well, existing bore well or sump available within the compound all the down-take pipes could be connected and led into it.

Percolation pit method   When there is a paved pathway, it is covered with perforated concrete slabs wherever necessary, to make more water to percolate down the soil.

Percolation pits are of 2–3 m deep, filled with broken bricks or pebbles with river sand on top (Figure 17.5). The rainwater gets filtered in the process of flowing through layers and reaches the underground aquifers and recharges it. In the case of clayey soils, percolation pit with bore well can be used.

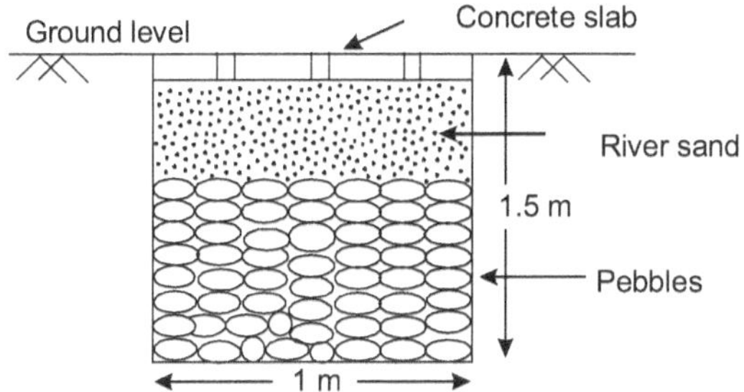

Figure 17.5   Percolation pit

Percolation pit with bore method   Whenever the depth of clay soil is more, recharge through percolation pit with bore is preferable. The bores

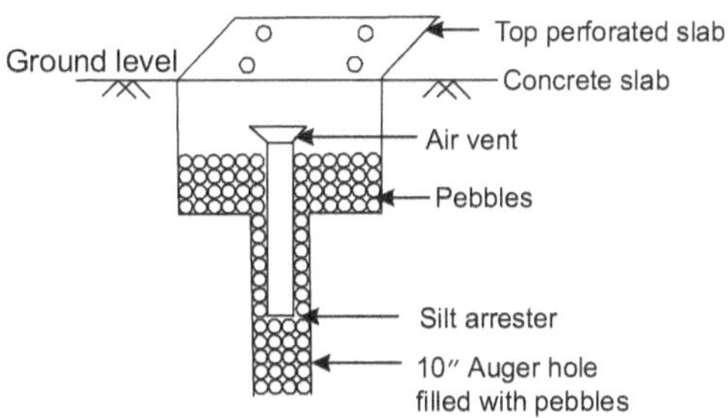

Figure 17.6   Percolation pit with bore

are deep pits about 20′ to 25′ deep depending on the soil condition. This bore can be at the centre of the square and is filled with pebbles or pieces of bricks and the top portion with river sand and covered with perforated concrete slab (Figure 17.6).

Recharge trench cum percolation pit (TCP) method   The TCP is one of the most ideal recharge structures in limited open land, narrow streets and roads, etc. to trap the surface water and recharge the aquifer. The dimensions of the TCP depend on the specific conditions of the site. One or two suitable shafts can also be provided inside the trench depending on the subsoil formation and the availability of water for recharge purposes. The entire structure may be filled with pebbles or locally available boulders and sand and if necessary this structure can be covered with perforated slabs.

The TCP can be located on sides of roads, street corners, low-lying areas inside the premises of individual houses, multi-storey buildings, paved areas surrounding the houses, near the extraction structures like bore wells, dug wells, etc.

Existing open well method   In the case of a dug well available within the house/flat complex, all the down-take pipes could be connected into it either through a filter or not, depending on the quality and quantity of water available in the said well.

Existing bore well method   In the case of a bore well available within the house/flat complex, all the down-take pipes could be connected into a filtering unit of size 21/2′ × 21/2′ × 21/2′ and the filtered water can be allowed into the existing bore well.

Storage tank method   If a sump or storage structure is available within the site, one or more down pipes could be directed into it through a sand

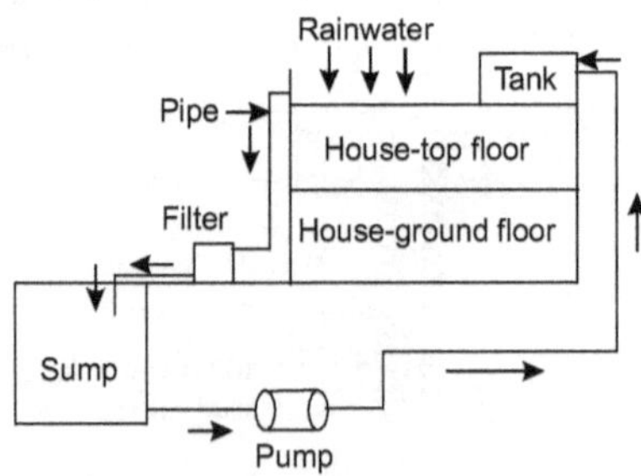

Figure 17.7   Storage tank

filter containing a layer of sand (sandwiched between two layers of pebbles/ blue metal) for immediate use (Figure 17.7). The overflowing water can be led into the well and in the absence of a dug well within the site, the down-take pipes could be led into either a bore pit or a percolation well depending on the volume of the rooftop water.

Surface Run-off Water Harvesting

Percolation ponds   Percolation ponds are small water storage structures constructed across natural streams and watercourses to harvest the run-off from the catchments and impound for a longer time to facilitate percolation of impounded water into the soil substrata both vertically and laterally, thereby recharging the groundwater.

Check-dams   Check-dams are small and moderate barriers constructed across a gully or stream to mellow down the velocity of water and to make the sediments carried by the water to settle down. Though check-dams are constructed as gully control measures, they serve as groundwater recharging structures.

Farm ponds   Dug out ponds on farm ponds can be constructed in a relatively flat terrain. The water is stored in the surface pond and surface withdrawal is possible for utilization. The low point of a natural depression is a good location for an excavated pond or farm pond.

Injection wells   Here the groundwater storage of a confined aquifer is done by "pumping in" treated surface water under pressure. Because of low permeability in the confining aquifer, it cannot get natural replenishment from the surface and needs direct injection through recharge wells. Injection wells are also used in coastal regions to arrest the ingress of seawater and to combat the problems of land subsidence in areas where confined aquifers are heavily pumped.

Spreading methods   The method comprises increase in surface area of infiltration to induce more infiltration. Spreading methods are suitable where large area for basin is available. The various spreading methods include in the following.

Flooding method   It is the cheapest method and is suitable for relatively flat areas where the topography is not spotted with large gullies ridges. The water is spread as a thin sheet of over the undisturbed native vegetation and soil cover for higher rate of filtration.

Basin method   In this method, water is improved in a series of basins formed by a network of small banks, check-dams and surface dykes. The basins are arranged so that the entire area is submerged during spreading operations to the most common method of artificial recharge, suitable for areas where the ground surface is irregular such as shallow gullies and ridges.

Channel method   It consists of a system of closely spaced flat-bottomed channels to expose the maximum infiltrating area suitable for irregular areas and in places where slopes are too steep for basin construction.

Pits and shafts   In this method, recharge is done through pits or shafts dug to puncture the impervious layer between the surface and water table in the area.

RESETTLEMENT AND REHABILITATION

People are disturbed due to severe economic, social and environmental problems listed below.

- Disturbance in the normal production.
- People are relocated to environments where their productive skills may be less applicable.
- Community structures and social networks are weakened.
- Relatives are dispersed.

The World Bank was the first multilateral lending agency that adopted a policy for Resettlement and Rehabilitation (R and R). The aims of R and R are as follows:

Aims of resettlement and rehabilitation policy

- Minimizing involuntary resettlement;
- Providing displaced people by a project with the means to improve, or at least restore, their former living standards, earning capacity and production levels;
- Involving both resettlers and hosts in resettlement activities;
- A time-bound resettlement plan; and
- Valuation and compensation principles for land and other assets affected by the project.

Few projects requiring R and R polices are

◇ Dams and reservoirs

◇ Industrial plants (large-scale) and industrial estates

◇ Land clearance and levelling

◇ Port and harbour development

◇ Resettlement and all projects with potentially major impacts on people

◇ River basin development

◇ Manufacture, transportation, and use of pesticides or other hazardous and/or toxic material

◇ Agro-industries (small-scale)

◇ Electrical transmission

◇ Irrigation and drainage (small-scale)

◇ Rural electrification

◇ Tourism

◇ Watershed projects (management or rehabilitation)

◇ Rehabilitation, maintenance, and upgrading projects (small-scale)

The objectives of this policy are the following.

◇ Particular attention should be paid to the needs of the poorest groups to be resettled.

◇ Community participation in planning and implementing resettlement should be encouraged. Existing social and cultural institutions of resettlers and their hosts should be supported and used to the greatest extent possible.

◇ Resettlers should be integrated socially and economically into host communities.

◇ Land, housing, infrastructure and other compensations should be provided to the adversely affected population, indigenous groups, and ethnic minorities.

The resettlement and rehabilitation plan should contain

◇ Organizational responsibilities

◇ Community participation and integration with host populations

- ◇ Socio-economic survey
- ◇ Legal framework
- ◇ Alternative sites and selection
- ◇ Valuation of and compensation for lost assets
- ◇ Land tenure, acquisition, and transfer
- ◇ Access to training, employment and credit
- ◇ Shelter, infrastructure, and social services
- ◇ Environmental protection and management
- ◇ Implementation schedule, monitoring and evaluation

RECLAMATION OF WASTELANDS

Wastelands include degraded forests, drought-struck pastures, eroded valleys, hilly slopes, waterlogged marshy lands, barren lands, etc.

Different authors have defined wastelands as follows:

- ◇ Land which is lying unproductive or which is not being utilized to its potential.
- ◇ Land which is incapable of producing materials or services of value (American Society of Soil Science).
- ◇ Land which has been abandoned and for which there is no further use (e.g. abandoned quarries or mine spoils).
- ◇ Land which produces less than 20% of the economic potential.
- ◇ Land where no greenery can be sustained.
- ◇ Lands which are ecologically unstable, badly eroded and degraded.
- ◇ Land which is either under forest cover or agricultural cover, or assigned for specific purposes such as national parks or national hydel projects.

The main causes of formation of wastelands are

- ◇ indiscriminate and over-utilization of forest produces
- ◇ over-grazing
- ◇ side effects of development projects
- ◇ misuse and unscientific land management

Classification of Wastelands

Wastelands are of two types, namely, culturable wastelands and unculturable wastelands.

Culturable wastelands   These are cultivable wastelands which are not being utilized to their full potential or are being mismanaged due to various reasons such as state or private occupation or having been declared as notified forest area. Such culturable wastelands include gullied lands, surface-waterlogged marshy lands, undulating and saline lands. Also included in this category are the wastelands based on ecological limitations such as degraded forests and pastures, shifting cultivation areas, sand dunes or mining spoils, etc.

Unculturable wastelands   These are the wastelands which are not available for cultivation. These include barren rocky lands, steep sloping areas and areas covered by snow or glaciers.

Significance of Wastelands

Formation of wastelands leads to the deterioration of the ecological balance by adversely affecting the various components of the ecosystem directly or indirectly dependent on that particular land.

Problems and Prospects of Wasteland Development in Tamil Nadu

After the green revolution such as extensive and intensive agriculture, the nation is now confronted by the second-generation production problems of the first green revolution, such as depletion of soil nutrients and water resources, creation of salinity and waterlogging, resurgence of pests and diseases, increased environmental pollution, factor productivity decline, etc.

Yet, agricultural production has to be increased to meet the demands of growing population and industry. Therefore, land use planning assumes greater significance on account of employment, agricultural production, income generation and sustainable development. At present, one of the options of increasing the agricultural production is increasing the area under cultivation. Bringing wastelands under cultivation could bring about expansion in area under cultivation.

Current Scenario-India

In India, wastelands accounted for about 20 per cent of the area surveyed (31. 66 million hectares) by the Institute of Remote Sensing. The land use classification and the wasteland classification of the National Remote Sensing Agency has classified as below:

The National Remote Sensing Agency (Department of Space), Hyderabad has generated wasteland maps on 1 : 50000 scale for different districts of Tamil Nadu. Among thirteen types of wastelands, the under-utilized/degraded notified forestlands account for 36.74 per cent, the uplands with or without scrub account for 36.00 per cent, and the lands affected by salinity and alkalinity account for 11.85 per cent. Altogether 84.59 per cent of the total area is accounted under wastelands in Tamil Nadu.

Though there are different ways of assessing the extent of wastelands, the High Power Committee on Development of Wastelands, Watershed Development and Implementation of Poverty Alleviation Programmes, has recommended that the extent and category of wasteland as per National Remote Sensing Agency classification and as determined by remote sensing techniques may be used when complete remote sensing data are available.

Lands get degraded on account of various reasons.

- ◇ Erosion by water and wind, chemical changes through accumulation of salts.
- ◇ Waterlogging, reckless felling of trees.
- ◇ Overgrazing in catchment areas of rivers is the major cause of soil degradation. Farmers for various reasons keep cultivable lands fallow.
- ◇ Non-profitable nature of farming, unsuitable terrain due to soil erosion, disputes and litigations.
- ◇ Problems in the management of cultivation are the important reasons for not cultivating all the lands.
- ◇ In Uttar Pradesh, lack of irrigation, soil problems, water stagnation, problem of accessibility (lack of link roads),

menace of wild animals, legal disputes, and financial problems are the major reasons for leaving the land fallow in lands need to be identified for developing strategies for their full utilization.

The factors responsible for the presence of fallow lands are as follows:

◇　natural endowment factors like rainfall pattern and the extent of dryland,

◇　factors representing area development pressures such as growth in urbanization, industrialization and infrastructures,

◇　factors representing sectoral development (within agriculture) like irrigation,

◇　mechanization, commercialization and institutional developments,

◇　personal factors like inability, litigation, resource base, etc.

The above factors have greatly damaged the productive capacity of land by reducing the infiltration, water retention capacity, increasing run-off, disrupting hydrological cycles, formation of gullies and ravines and reducing soil fertility.

The task of reclaiming wastelands is a complex one since each category of wasteland has its own causative origin and needs special approach for reclamation, taking into account the rainfall, soil and topography in which the wastelands are located.

A Holistic approach integrating technologies must be pursued in complete form rather than a piecemeal approach to redeem the wastelands.

Any programme on wasteland development must consider the following two objectives; (i) identification of the nature, extent and location of wastelands and their availability for agricultural uses and (ii) identification of technology and management practices suitable for specific type of wasteland taking into account the availability of resources for their care after planting or introduction.

Wasteland development has been stated off in Tamil Nadu as in the name of PILOT in 10 districts, by the State Department of Agriculture.

Wasteland development programme includes farmers' training and starting of nurseries. The work has just started and will take 5 years to cover all the watersheds. The total outlay for this 55,000 hectare project has been estimated at Rs. 70 crores. The government had extended Rs. 25 crores during the first year.

A full-scale project is on the cards, on an estimated outlay of Rs. 1500 crores. World Bank and other agencies are supposed to give financial support to this programme.

Climate change, also called global warming, refers to the long-term fluctuation in temperature, precipitation, wind and other elements of the Earth's climate system. Already, findings so far suggest that the earth's temperature has risen by 0.3–0.6°C since the late 19th century.

- ⋄ Global sea level has risen by 10–25 cm over the past 100 years.
- ⋄ It is expected to rise between 9 cm and 29 cm by 2030 and 28 cm and 96 cm by 2090.
- ⋄ Research by the International Rice Research Institute, Manila, has indicated that every 1°C rise in temperature will result in a 10 per cent fall in the yield of rice.

NATIONAL POLICY FOR CLIMATE CHANGE MITIGATION

The Ministry of Environment and Forests is the nodal Ministry for all environment-related activities in the country and for coordinating the climate change policy as well. The working group on the FCCC (Framework Convention of Climate Change) was constituted to oversee the implementation of obligations under the FCCC and to act as a consultative mechanism in the Government for impacts to policy formulation on climate change. To enlarge the feedback mechanism, the Government of India has constituted an Advisory group on climate change under the chairmanship of the Ministry of Environment and Forests. The policy of the Government of India on reduction of GHG (Greenhouse gas) emission is based on three broad principles, namely,

1. The primary responsibility of reducing green house gases (GHG) emission is that of developed countries, and hence should show a demonstrable sincerity in initiating actions to address climate change;

2. The developmental needs of developing countries are of prime importance; and

3. The developed countries should transfer resources and technologies at favourable terms to the developing countries, thereby facilitating developing countries to move towards a sustainable development path.

INTERGOVERNMENTAL PANEL ON CLIMATE CHANGE

The Intergovernmental Panel on Climate Change (IPCC) established jointly by the World Meteorological Organization (WMO) and the United Nations Environmental Program (UNEP), in 1988, has been mandated to assess all available factual information on the science, the impacts, and the economics of climate change and on the adaptation/mitigation options to address climate change.

Since the Earth Summit in Rio in 1992, it has been a long march for the world to reach a consensus and to commit together, on the road to action in combating global warming. The first international approach to climate change had taken shape with the development of the United Nations Framework Convention on Climate Change (UNFCCC). Adopted in 1992, the UNFCCC sets a framework for action aimed at stabilization of greenhouse gas concentrations, in the atmosphere at a level that would prevent dangerous anthropogenic interference with the climate system. The first Conference of the Parties to the Convention (COP-1) was held in Berlin in March–April, 1995. During the COP-1, it became clear to the world community that "Rio is not enough" and pressed for adequacy of commitment by the developed countries for reduction of GHGs emissions. Accordingly, a Protocol to the Climate Change Convention was adopted in Kyoto in 1997, now known as the Kyoto Protocol. The objective of the Kyoto Protocol is aimed at bringing down the global GHG emissions by 5.2% during the year 2008–2012. The IPCC estimates that the global mean surface temperature would be 2°C above the pre-industrial levels by the year 2030, and about 4°C above pre-industrial levels by the year 2090.

Ever since the inception of the UNFCCC in 1992, the Government of India has been very active in the climate change negotiations.

GREENHOUSE EFFECT

The greenhouse effect is primarily a function of the concentration of water vapour, carbon dioxide, and other trace gases in the atmosphere that absorb the terrestrial radiation leaving the surface of the Earth, and act like a blanket over the earth's surface, keeping it warmer than it would otherwise be (Figure 18.1). Changes in the atmospheric concentration of these gases can alter the balance of energy transfers between the atmosphere, space, land, and the oceans. A gauge of these changes is called the radiative forcing, which is a measure of changes in the energy available to the earth-atmosphere system. Holding everything constant, increases in greenhouse gas concentrations in the atmosphere will produce positive radiative forcing (i.e., net increase in the absorption of energy by the earth).

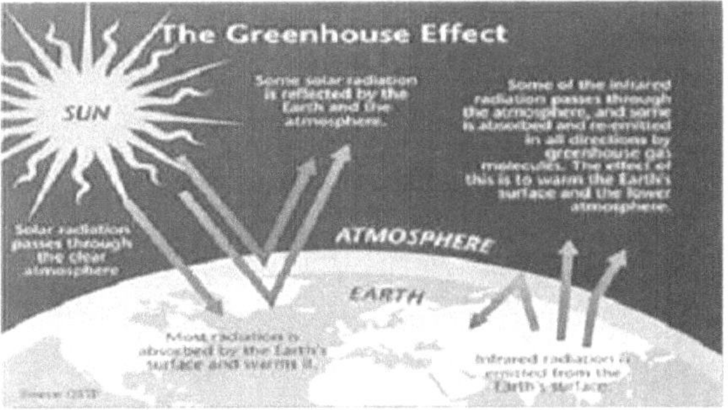

Figure 18.1   The greenhouse effect

Greenhouse Gases

The Earth's atmosphere primarily consists of oxygen and nitrogen, but neither plays any significant role in the greenhouse effect, as both are essentially transparent to terrestrial radiation. The gases currently known to cause the greenhouse effects include carbon dioxide (CO_2), methane (CH_4), nitrous oxide (N_2O), chlorofluorocarbons (CFCs), and two CFC substitutes, hydrochlorofluorocarbon (HCFC-22), and perfluoromethane (CF_4). Of these, carbon dioxide, methane, nitrous oxide and ozone are the naturally occurring greenhouse gases. Certain human activities however add to the levels of most of these naturally occurring gases. Carbon dioxide is responsible for over half the enhancement of the greenhouse effect (Figure 18.2).

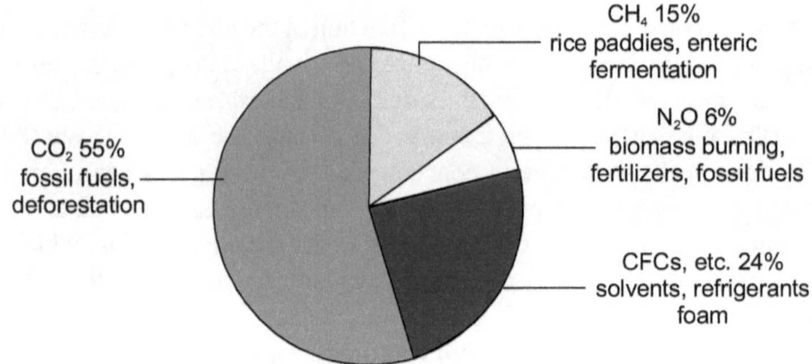

Figure 18.2   Greenhouse gases to global warming and their sources (1980–1990)

There are three main sources of greenhouse gases

◇ *Solar sources*   Changes in solar constant, XUV region, solar proton event and changes in ozone content generate the trace gases.

◇ *Volcanic sources* Cl_x and a number of molecular species like CO_2, H_2, CH_4, CO_3 are of greenhouse interest. Mt. Agung alone that erupted in 1963 spewed 3×10^{32} molecules.

◇ *Industrial sources* CO_2, N_2O, CH_2 and CFCs are derived from biospheric or industrial sources.

CO₂ (Carbon dioxide)   The biosphere annually produces 194 gt of carbon (1 gigatons = 1000 million tons). The biosphere is doing its best to cope with the extra burden but can retain only half of this extra carbon. As a result, 17 ppm/v (parts per million by volume) of CO_2 is added every year. There is no doubt that atmospheric CO_2 levels are arising dramatically and half of this rise has occurred since 1960.

The best known of the greenhouse gases is CO_2, which is emitted into the atmosphere in large amounts the equivalent of more than 6,000 tons of carbon per year. The industrialized countries particularly the US are largely responsible for the accumulation of greenhouse gases in the atmosphere.

CO$_2$ concentration has been increasing at a very rapid rate and has now reached a level of 353 ppm/v.

The contribution made by India to global concentration of CO$_2$ amounted to 140 million tons in 1988–89. In other words, 2.8 per cent global CO$_2$ warming was caused by India. Figure 18.3 shows the emission of CO$_2$ in different countries.

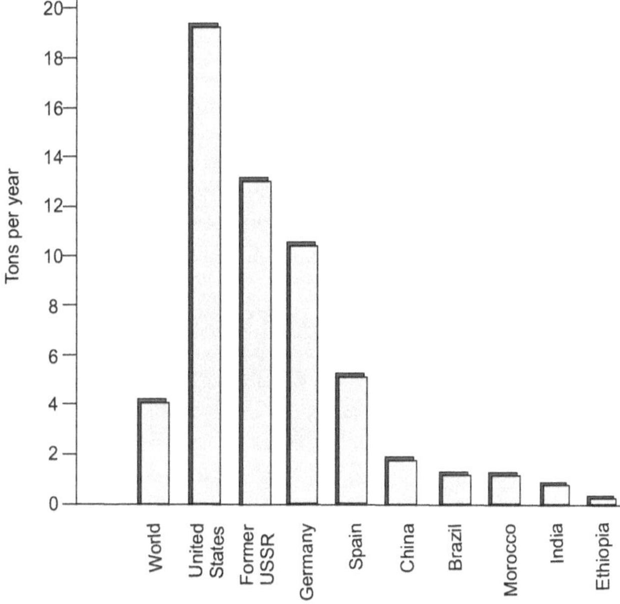

Figure 18.3   CO$_2$ emission in different countries

N$_2$O (Nitrous oxide)   Combustion of coal, oil, biomass and fertilizers, modern waste disposal practices, grasslands, woodlands and tropical and subtropical forests largely contribute to atmospheric N$_2$O. It is increasing at the rate of 0.7 ppb/year.

CH$_4$ (Methane)   According to an estimate, total concentration of methane was 1630 ppb in 1984. The main sources are sheep and cattle dung, rice-paddy fields, wetlands, burning of the biomass and mining of coal and mineral oil. CH$_4$ is 25 times more effective than CO$_2$ in absorbing the outgoing infrared radiation.

CFCs (Chlorofluorocarbons) Chlorofluorocarbons do not occur naturally and are totally man-made. They are used in refrigerators, air conditioners, as aerosol propellants and cleaning agents. India produces about 7000 tons of CFCs out of the global annual production of over one million tons. India is one of the signatories of the Montreal Protocol. Montreal Protocol calls for a 50 per cent reduction in the use of chlorofluorocarbons by 1998. CFCs, halons and other ozone depletors can be abandoned without which substitutes themselves must be environmentally safe.

India is one of the few developing nations which consumes as well produces ozone depleting substances (ODS). In fact, next to China, India is the largest producer of chlorofluorocarbons.

Companies making ODS-based products have to stop producing these substances. Production of CFC-based product is banned in India from January 1, 2003 while their own phase out date is fixed 2010.

CLIMATE CHANGE IMPACTS

◇   The global average temperature is projected by models to increase by 1.4–5.8°C by 2100 relative to 1990. It also concludes that the global average sea level is projected by models to rise 0.09–0.88 m by 2100. These projections indicate that the warming would vary by region, and be accompanied by increases and decreases in precipitation.

◇   In addition there would be changes in the variability of climate, and changes in the frequency and intensity of some climate phenomena.

◇   Many human-associated systems sensitive to climate change include water resources, agriculture and forestry; coastal zones and marine system; human settlements, energy, and industry; insurance and other financial services, and human health.

◇   The vulnerability of these systems varies with geographical location, time, and social, economical, and environmental conditions. Natural systems are especially vulnerable to climate change because of limited adaptive capacity. Some of these systems may undergo significant and irreversible damages.

◇   Natural systems at risk include glaciers, coral reefs, mangroves, tropical forests, polar and alpine ecosystems, etc.

◇ While some species may increase in abundance or range, climate change will increase the existing risk of extinction of some more vulnerable species and loss of biodiversity.

◇ The vulnerability of human societies and natural systems to climate extremes is demonstrated by the damage, hardship, and death caused by events such as droughts, floods, heat waves, avalanches, and windstorms.

◇ The effects of climate change are expected to be the greatest in the developing countries in terms of loss of life and relative effects on investment and the economy. Many regions that are vulnerable to climate change are also under pressure from forces such as population growth, resource depletion, and poverty.

Effects on India

◇ India could be more at risk than many other countries from changes in temperature and sea level. Models predict an average increase in temperature in India of 2.3–4.8°C for the benchmark doubling of carbon dioxide scenario.

◇ Temperatures would rise more in northern India than in southern India. In the North Indian Ocean, under a doubling, the average number of tropical disturbance days could increase from 17–29 a year.

◇ While, without protection, approximately 7 million people would be displaced, and 5,760 km^2 of land and 4,200 km of road would be lost. The dominant cost would be the land loss, accounting for 83 per cent of all damages.

For the same CO_2 doubling scenario, a crop simulation study estimates that wheat yields could decrease between 28 and 68 per cent. Even allowing for adaptation options, agricultural losses could be significant. The loss in farm revenues is estimated at 9–25 per cent for a temperature rise of 2–30°C.

Human Consequences of these Changes

◇ 5–50 per cent increase in drought frequencies.

◇ 15 per cent destruction of the arable land in countries like Egypt and Bangladesh.

◇ Nations like Maldives and Kiribati are much affected by rising sea level.

◇ Millions of environmental refugees.

Climate Change on Ecological Socio-economic System

Climate change represents an additional stress on the ecological and socio-economic system. A study conducted jointly by the Indian Agricultural Research Institute, New Delhi and Central Institute of Dryland Agriculture, Hyderabad, stated that in the current scenario of global climate change, the effect of climate change on "kharif crop" is expected to be less than that on "rabi crop". Rainfall in rabi season will, however, have wider uncertainty. Kharif rainfall is likely to increase by as much as 10 per cent. Further, the study stated that the onset of monsoon over India is projected to be delayed and often uncertain.

In 1999, tropical cyclone that hit Orissa causing a death toll of about 10,000, clearly demonstrates the extreme significance of impacts related to climate variability. The low-lying and densely populated Indian coastline extending to about 6500 kms remains highly vulnerable to any sea level rise.

◇ Ecosystems will generally shift northward or upward in altitude, but in some cases they will run out of space, as a 1°C change in temperature corresponds to a 100 km change in latitude. Hence, average shift in habitat conditions by the year 2100 will be of the order of 140–580 km.

◇ Pollination may be disrupted—higher temperatures and earlier snowmelt may trigger earlier flowering. This could affect interactions with other species that depend on flowering plants. For example, alteration of the distribution and growing season of the Upland Larkspur could adversely impact pollinators such as broad-tailed hummingbirds and bumblebees.

◇ Bird species with limited ranges may be lost. For example, the Kirtland Warbler nests in northern Michigan, whose habitat is the declining jack pines in the region.

◇ Melting of polar ice may cause problems for polar bears.

◇ Coral reef mortality may increase and erosion may be accelerated. In 1998 alone, 16% of the world's coral reef died from higher temperatures. Increased level of carbon dioxide adversely impacts the coral building process (calcification). Scientists estimate that calcification could decline by 17–35% below pre-industrial levels by 2100.

◇ Sea level may rise, engulfing low-lying areas. This causes disappearance of many islands, and extinctions of endemic island species.

◇ The number of whales may decline. A significant reduction in sea ice could adversely affect the abundance of krill, the primary source of food for whales in the southern hemisphere.

◇ Invasive species may be aided by climate change. Exotic species can outcompete native wildlife for space, food, water and other resources, and may also prey on native wildlife.

◇ Droughts and wildfires may increase. An increased risk of wildfires due to warming and drying out of vegetation is likely.

◇ Sustained climate change may change the competitive balance among species and might lead to forests dieback.

◇ Climate-change-induced droughts may push humans into wildlife habitat.

Energy released during a nuclear reaction in accordance with mass-energy equation is called "nuclear energy". A small quantity of radioactive material can produce an enormous amount of energy. For instance, one ton of uranium would provide as much energy as 3 million tons of coal or 12 million barrels of oil.

The first nuclear power plant in the world, designed mainly for electricity production, came into operation in 1957 at Calder Hall in Cumberland district of the United Kingdom. At present there are over 300 atomic power plants in USA (83), followed by former USSR (40), UK (35), France (34), Japan (25), Germany (15) and Canada (13).

Till date nuclear accidents of the Three Mile Island and Chernobyl reactors are the two most well-known accidents.

RADIATION EFFECT

◇ Radiations may break the chemical bonds such as DNA in cells. This affects the genetic make-up and control mechanisms. The effects may be instantaneous, prolonged or delayed types. It could even be carried to future generations.

◇ When exposed to low doses of radiations (100–250 rads), men do not die but begin to suffer from fatigue, nausea, vomiting and loss of hair.

◇ If exposed to higher doses (400–500 rads), the bone marrow is affected, and blood cells are reduced.

◇ Higher irradiation doses (10,000 rads) kill the organisms by damaging the tissues of heart, brain, etc.

◇ Through the food chain also, radioactivity effects are experienced by humans.

Most information concerning the effects of high doses of radiation on people are from studies of the people who survived the atomic bomb detonations in Japan at the end of World War II.

Recent Nuclear Disasters

Year	Disaster
October 7, 1957	A fire in the wind scale plutonium production reactor N of Liverpool, England, spread radioactive material throughout the countryside. In 1983, the British government said that 39 people probably died of cancer as a result.
January 3, 1961	An experimental reactor at a federal installation near Idaho Falls, Id., killed three workers—the only deaths in US reactor operations. The plant had high-radiation levels but damage was contained.
October 5, 1966	A sodium cooling system malfunction caused a partial core meltdown at the Enrico Fermi demonstration breeder reactor near Detroit, Mich. Radiation was contained.
January 21, 1969	A coolant malfunction from an experimental underground reactor at Lucens Vad, Switzerland, resulted in the release of a large amount of radiation into a cavern, which was then sealed.
March 22, 1975	A technician checking for air leaks with a lighted candle caused a $100 million fire at the Brown's Ferry reactor in Decatur, Ala. The fire burned out electrical controls, lowering the cooling water to dangerous levels.
March 28, 1979	The worst commercial nuclear accident in the US occurred as equipment failures and human mistakes led to a loss of coolant and partial core meltdown at the *Three Mile Island* reactor in Middletown, Pa. Thousands living near the plant left the area before the 12-day crisis ended, during which time some radioactive water and gases were released.
February 11, 1981	Eight workers were contaminated when over 100,000 gallons of radioactive coolant leaked into the containment building of the TVA's Sequoyah 1 plant in Tennessee.
April 25, 1981	Some 100 workers were exposed to radioactive material during repairs of a nuclear plant at Tsuruga, Japan.

Year	Disaster
January 6, 1986	A cylinder of nuclear material burst after being improperly heated at a KerrMcGee plant at Gore, Okla. One worker died and 100 were hospitalized.
April 26, 1986	In the worst accident in the history of the nuclear power industry, fires and explosions resulting from an unauthorized experiment at the Chernobyl nuclear power plant near Kiev, USSR (now in Ukraine), left at least 31 people dead in the immediate aftermath of the disaster and spread significant quantities of radioactive material over much of Europe. An estimated 135,000 people were evacuated from area around Chernobyl, some of which were rendered uninhabitable for years. As a result of the radiation released into the atmosphere, tens of thousands of excess cancer deaths (as well as increased rates of birth defects) were expected in succeeding decades.
March 24, 1992	At the Sosnovy Bor station near St. Petersburg, Russia, radioactive iodine escaped into the atmosphere. A loss of pressure in a reactor channel was the source of the accident.
November 1992	In France's most serious nuclear accident, three workers were contaminated after entering a nuclear particle accelerator in Forbach without protective clothing. Executives were jailed in 1993 for failing to take proper safety measures.
November 1995	Japan's Monju prototype fast-breeder nuclear reactor leaked two to three tons of sodium from the reactor's secondary cooling system.
March 1997	The state-run Power Reactor and Nuclear Fuel Development Corporation reprocessing plant at Tokaimura, Japan, contaminated at least 35 workers with minor radiation after a fire and explosion occurred.
September 30, 1999	Another accident at the uranium-processing plant at Tokaimura, Japan, plant exposed fifty-five workers to radiation. More than 300,000 people living near the plant were ordered to stay indoors. Workers had been mixing uranium with nitric acid to make nuclear fuel, but had used too much uranium and set off the accidental uncontrolled reaction.

Year	Disaster
August 12, 2000	Explosions on board a Russian nuclear powered submarine caused the vessel to sink in the Barents Sea, killing all 118 on the board and hitting the seabed. The submarine was recovered and defuelled by February 17th, 2003.
September, 2000	Two to five people were contaminated with radiation from an accident at a nuclear power plant in Bulgaria.
Winter, 2002	A shipment of radioactive iridium-192 leaked on its way from Sweden to the USA. The package leaked quickly, however it is not known when or where the leak began, and at what level, people were exposed to radioactivity.
October 23, 2002	A nuclear-powered submarine caught fire on the deck and then the rubber coating. The fire did not cause any release of radiation into the environment.
March 5, 2003	A fire broke out on a nuclear submarine in Russia causing damage to the coating of the submarine, although no radiation leakage was reported.
April 10, 2003	A leakage of radioactive gases occurred during cleaning at the Paks nuclear power plant, Hungary. The incident is rated at level three because the gases were released into the environment.

Case Study 18

Bhopal is the site of the greatest industrial disaster in history. On the night of December 2nd, 1984, a dangerous chemical reaction occurred in the Union Carbide factory when a large amount of water got into the MIC (methyl isocyanate) storage tank. The leak was first detected by workers about 11:30 p.m. when their eyes began to tear and burn. They informed their supervisor who failed to take action until it was too late. In that time, a large amount, about 40 tons of MIC, poured out of the tank for nearly two hours and escaped into the air, spreading within eight kilometres downwind, over the city of nearly 900,000 people. Thousands of people were killed (estimates ranging as high as 4,000) in their sleep or as they fled in terror, and hundreds of thousands remain injured or affected (estimates range as high as 400,000) to this day. The most seriously affected areas were the densely populated shanty towns immediately surrounding the plant. The victims were almost entirely the poorest members of the population.

This poisonous gas, caused death and left the survivors with lingering disability and diseases. Not much is known about the future medical damage of MIC, but according to an international medical commission, the victims suffer from serious health problems that are being misdiagnosed or ignored by local doctors. Exposure to MIC has resulted in damage to the eyes and lungs and has caused respiratory ailments such as chronic bronchitis and emphysema, gastrointestinal problems like hyperacidity and chronic gastritis, ophthalmic problems like chronic conjunctivitis and early cataracts, vision problems, neurological disorders such as memory and motor skills, psychiatric problems of various types including varying grades of anxiety and depression, musculoskeletal problems and gynaecological problems among the victims. It is estimated that children born in Bhopal after the disaster face twice the risk of dying as do children elsewhere, partly because parents cannot care for them adequately. Surprisingly enough, despite the serious health problems and the deaths that have occurred, Union Carbide claims that the MIC is merely a "mild throat and ear irritant".

The Bhopal disaster was the result of a combination of legal, technological, organizational, and human errors. The immediate cause of the chemical reaction was the seepage of water (500 litres) into the MIC storage tank. The results of this reaction were exacerbated by the failure of containment and safety measures and by a complete absence of community information and emergency procedures. The long-term effects were made worse by the absence of systems to care for and compensate the victims.

The Continuing Holocaust—Bhopal

Every month in Bhopal at least four to five protests are organized on the ongoing issues of the Bhopal disaster. For the survivors in Bhopal, the disaster is still continuing.

More than 20,000 people living behind the Union Carbide factory are still forced to drink water laced with dichlorobenzene, trichlorobenzene, chloroform and heavy metals such as lead mercury. To this day, 15 to 20 people die every month due to exposure-related illnesses. There are more than 1,50,000 chronically sick people, too ill to work, who have never been compensated for their loss of work. There are children in the second generation whose growths have been stunted because their parents were exposed.

During the age of hunting and gathering, the population was less than 1 million, and the average growth rate was about 1 per 200 sq.km having a death rate less than 0.00011% per year. During agricultural revolution, the population increased to 1 or 2 people/sq.km and the population was 100 million in 1 AD. During the industrial revolution, the total human population was about 900 millions in 1800 AD and doubled to 3 billions by 1960. The main reasons for this growth are discovery of vaccines, sufficiency in production of food, shelter and clothing.

Currently the world's population is growing exponentially at a rate of about 1.26% per year. The ticking of this population clock means that in 2003 the world's population of 6.3 billion grew by 79 million people (6.3 billion × 0.0126 = 79 million), an average increase of 216,000 people a day, i.e., 9,000 an hour. Each second, 79 million new babies are born. Spending 1 second saying hello to each of 79 million new people added to the earth during 2003 for 24 hours a day would take you about 2.5 years. By then there would be about 198 million more people to shake hands with. Figure 20.1 shows the increase in population growth.

Figure 20.2 shows the average annual increase in the world's population and the increase in population size from 1950 to 2003 and the projected increases to 2050. The six nations expected to experience most of this growth are India, China, Pakistan, Nigeria, Bangladesh, and Indonesia. The distribution of population depends to a large extent on the quality of land. Thus population density varies widely. Some of the densely populated areas are Western Europe, the Indian subcontinent, the plains and rivers, Nile valley, and also in the valleys of China and Northeastern USA.

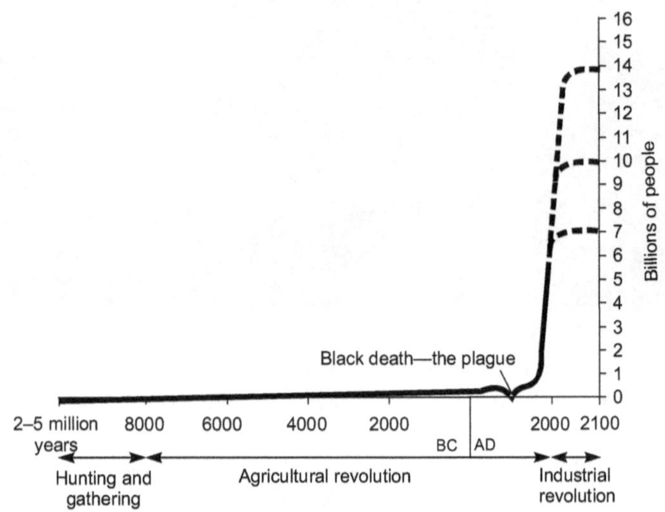

Figure 20.1   Graph showing an increase in population growth

Control measures should be implemented successfully or else the biosphere will collapse under the weight of the rapidly growing population.

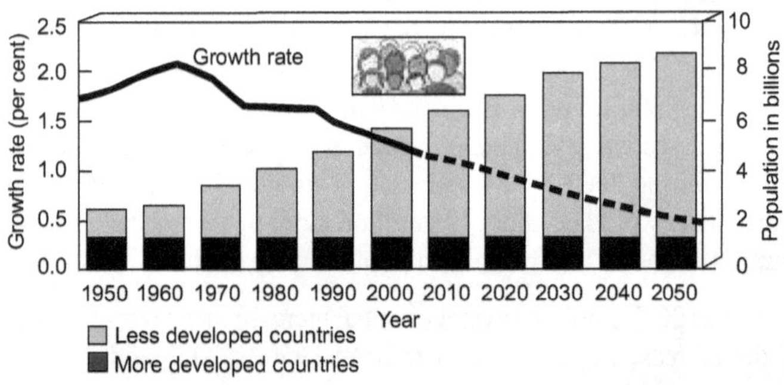

Figure 20.2   Average annual increase in the world's population

The factors encouraging settlement are good climate for a wide range of crops, resources, roads, railways development, etc.

The rate of growth has slowed down in developed countries, but population still continues to grow rapidly in developing countries.

The average rate of growth is 2% in the 20th century and is declined to 1.3% in 2002 AD. By 2025, the world's population will be 8 billions, and 9 billions by 2050. An increase of 9.5% is contributed by developing countries.

STABILIZATION OF POPULATION GROWTH

In 1994, the United Nations held its third once-in-a-decade Conference on Population and Development in Cairo, Egypt. The main aim of this conference was to encourage action to stabilize the world's population at 7.8 billion by 2050 instead of the projected 9.3 billion. By 2015, the major goals of the resulting population plan, endorsed by 180 governments, are

- to provide a universal access to family-planning and services, and reproductive health care.

- to improve the health care of infants, children, and pregnant women.

- to encourage development and implementation of national population policies as part of social and economic development policies.

- to bring about more equitable relationships between men and women, with emphasis on improving the status of women and expanding education and job opportunities for young women.

- to increase access to education, especially for girls and women.

- to increase the involvement often in child-rearing responsibilities and family planning.

- to take steps to eradicate poverty.

- to reduce and eliminate unsustainable patterns of production and consumption.

The experiences of Japan, Thailand, South Korea, Taiwan, Iran, and China indicate that a country can achieve or come close to replacement-level fertility within 15–30 years. Such experience also suggests that the best way to slow down the population growth is a combination of investing in family planning, reducing poverty, and elevating the status of women.

ECONOMIC REWARDS AND PENALTIES TO REDUCE BIRTH RATES

Some population experts point out that most couples in developing countries want three or four children. This is well above the replacement-level fertility that is needed to slow or halt a country's population growth.

These analysts believe that we must go beyond family planning and offer economic rewards and penalties to help slow population growth. About 20 countries offer small payments to people who agree to use contraceptives or to be sterilized. However, such payments are most likely to attract people who already have all the children they want.

Some countries, including China, penalize the couples, who have more than one or two children, by raising their taxes, charging other fees, or eliminating income tax deductions for a couple's third child (as is done in Singapore, Hong Kong, and Ghana). Families who have more children than the prescribed limit may also lose health-care benefits, food allotments, and job options. Researches show that economic rewards and penalties designed to lower birth rates work best if they encourage (rather than coerce) people to have fewer children, reinforce existing customs and trends toward smaller families, do not penalize people who produced large families before the programmes were established, and improve a poor family's economic status. Once a country's population growth is out of control, it may be forced to use coercive methods to prevent mass starvation and hardship, as has been the case for China.

Population Explosion and Human Ecology

For decades the Malthusian Specter, "the world's population outstripping the available quantity of foodgrains" has kept haunting humanity. The fear has more or less proved to be unfounded in developed nations where the population growth has resulted in fullest utilization of natural resources. However, in developing countries such as India, the population explosion is eating up everything starting from natural resources and economic development to ecological balance.

Population explosion is a sudden spurt in the rate of population growth with improvement of living standards, expansion of education and reduction in death rate and overall development of a nation. Thus population increases at an alarming rate. Population explosion does not mean over-population. Economic development fails to maintain pace with population growth. Some countries like South Africa could not feed for the population. Japan has high population density but it ranks first in human development index.

Some 70 to 75 per cent of the earth's population is contained within four zones, two in Asia, one in Europe and one in Anglo-America together. The two Asiatic agglomerations account for more than 46 per cent of the world's inhabitants. India lies in the South Asian cluster with distinctly tropical climate. Climate is most suited for living and increasing fertility. Here the population presses closely upon the food supply, poverty and malnutrition are omnipresent, birth rates are still depressingly high, while death rate, although still high by western standards, are falling. Consequently, the rate of natural increase in pollutants is large and of course in absolute numbers is tremendous. Population explosion in India is putting on severe burden on natural resources like land, forest, mineral, water, fisheries, energy, etc.

The present world population figures show that there are 27 persons inhabiting per sq.km of area, which may not take a long time to increase to 1000 persons which would be a beyond the carrying capacity per sq.km. And if the present rate is to continue, probably in another thousand years people will occupy every space on the earth's surface including the hydrosphere. The scientific man will build dwelling places towards the sky and on the surface of the oceans. But finding some space to dwell is not the solution of the problem. The more urgent problems will be the availability of proper living environment.

In our country, the rapidly increasing population is making the natural resources availability scenario grim. The ecological pyramid is being disturbed and the abiotic environment is gradually degrading. If proper steps regarding the long-term policy of natural resources exploitation on the one hand and population control on the other is not adopted, the country will have to face grave consequences in the near future. We must take lesson from China's population policy which states, "We do not believe in anarchy in material production, and we do not believe in anarchy in human reproduction. Man must control nature and he must also control his numbers."

CONSEQUENCES OF POPULATION GROWTH

Population growth is one of the prominent factors that affects the degradation of the environment. Earlier, pestilence and famine kept the population under control, but with the development of chemical compounds to restore and enhance the soil fertility, and with reduction in the death rate, there has been an explosive growth in population with inevitable consequences.

Increased population growth leads to several consequences in the environment.

- unemployment
- low living standards of people
- hindrance in the process of development of economy, pressure on agricultural land
- low per capita income
- lack of basic amenities like water supply and sanitation education, health, etc.
- high-crime rate
- environmental damage
- migration to urban area in search of job
- energy crisis
- overcrowding of cities leading to development of slums.

Case Study 19

Slowing Population Growth in India

The world's first national family planning programme began in India in 1952, when its population was nearly 400 million. In 2003, after 51 years of population control efforts, India was the world's second most populous country, with a population of 1.1 billion.

In 1952, India added 5 million people to its population. In 2003, it added 17 million. India is projected to have a larger population than China by 2050, even if China's one-child policy is relaxed.

India faces a number of already serious problems like poverty, malnutrition, and environmental problems that could worsen as its population continues to grow rapidly. By global standards, India's people are poor. Nearly half of India's labour force is unemployed or can find only occasional work.

Some good news is that India currently is self-sufficient in food grain production. The bad news is that about 40% of its population and 53% of its children suffer from malnutrition, mostly because of poverty.

India faces some serious resource and environmental problems. With 17% of the world's people, it has just 2.3% of the world's land resources and 2% of the world's forests. About half of the country's cropland is degraded as a result of soil erosion, waterlogging, salinization, overgrazing, and deforestation. In addition, about 70% of India's water is seriously polluted and sanitation services often are inadequate.

Without its long-standing family planning programme, India's population and environmental problems would be growing even faster. Still, to its supporters the results of the programme have been disappointing because of poor planning, bureaucratic inefficiency, the low status of women (despite constitutional guarantees of equality), extreme poverty, and lack of administrative and financial support.

The government has provided information about the advantages of small families for years. Yet Indian women still have an average of 3.1 children. One reason is that most poor couples believe they need many children to do work and care for them in old age. Another is the strong cultural preference for male children, which means some couples keep having children until they produce one or more boys. These factors in part explain why even though 90% of Indian couples know of at least one modern birth control method, only 64% actually use one.

FAMILY WELFARE PROGRAMME

This programme aims at reducing the birth rate to the extent necessary to stabilize the population at a level consistent with the requirement of the national economy. In 1952, India launched a nationwide family planning programme making it the first country in the world to do so, giving a "Clinical approach" to the problem. However, during the Third Five Year Plan (1961–66), family planning was declared as the very centre of planned development. Extension education approach was followed for motivating the people for acceptance of the small family norm (Department of Family Planning in 1966 in the Ministry of Health). In the fourth Five Year Plan

(1964–74), the Government of India gave top priority to the programmes such as Mother and Child Health (MCH) activities of Primary Health Centres (PHCs). All India Hospital Postpartum Programme in 1970 and the Medical Termination of Pregnancy (MTP) in 1972 were introduced. During Fifth Five Year Plan (1975–80) there have been major changes. The country framed its first "National Population Policy" in 1976. The Ministry of Family Planning was renamed as "Family Welfare".

The salient features of the National Population Policy of 1986 are:

◊   advancing the age of marriage of girls to 20 years through intensified publicity campaigns and amendments in law.

◊   promoting the two-child family norm.

◊   increasing female literacy rate.

◊   promotion of spacing methods.   .

◊   enhancing child survival through universal immunization.

◊   revamping the infrastructure and improving programme management at all levels.

◊   raising a cadre of Women Volunteer Corporation.

Presently, the Family Welfare Programme seeks to promote on a voluntary basis, responsible and Planned Parenthood with one child norm, male or female through independent choice of family welfare methods best suited to acceptors. Women are forced to accept family planning methods. This shows that family planning has largely remained a woman-centred programme.

The instance of female infanticide in Punjab, Haryana, Sikkim, Uttar Pradesh, Delhi and Tamil Nadu and other states has brought down the sex ratio to an all-time low against females. Though there is a law banning the determination of the sex of the child in the womb, unscrupulous medical practitioners and short-sighted parents connive to prevent the birth of female children. There is, thus, an urgent need to prevent the misuse of technology through education and awareness. Though social-economic and education factors play a vital role in the population control programme; still economic empowerment of women can become an important factor in family welfare.

Family Welfare Programme is being implemented in Tamil Nadu to stabilize the population growth by improving the Maternal and Child Health

Services. Tamil Nadu is the model state in the introduction of Community Assessment Approach all over India. Family Welfare Programme was implemented as a peoples' programme involving the active cooperation of the community at large. The achievement of Family Welfare Programme in the state has been attributed to several factors including strong social and political commitment and good administrative backup.

Demographic Indicators

As per the 2001 census, the population of Tamil Nadu was 6.24 crores with a decadal growth rate of 11.7 per cent which is the second lowest in the country next to Kerala. Tamil Nadu is the sixth most populous state in India. It accounts for 6% of the country's total population. The population of Tamil Nadu for 2004 is estimated as 6.4 crores.

State Commission on Population

The Government of Tamil Nadu has constituted the "State Commission on Population" under the Chairmanship of Hon'ble Chief Minister of Tamil Nadu. The Commission will review, monitor and give directions in the implementation of family welfare and maternal and child health programmes in this state with a view to meeting the set goals. The Population Policy of the State will be published soon.

Insurance Scheme for Mothers who Undergo Sterilization

A unique scheme for providing insurance coverage for the mothers who undergo sterilization has been introduced from February 2001 onwards. Under this scheme, each mother will be insured by the government paying a premium of Rs.14/- per case. In the event of death due to sterilization in any government hospital/local body, voluntary organizations or approved nursing homes, a compensation of Rs. 2 lakhs will be paid to the family of the deceased mothers and deposited in the name of their children. For death outside the hospital within 30 days of discharge, a compensation of Rs. 25,000 will be paid to the family of the deceased mothers and deposited in the name of the children. The scheme is being implemented through the National Insurance Company Limited.

ENVIRONMENT AND HUMAN HEALTH

Environment has a direct impact on physical, mental and social well-being of those living in it. The environmental factors that impair human health are:

    ◇   pathogenic agents and their vectors and reservoirs

    ◇   physical, chemical and biological agents present in the environment

    ◇   noxious physical, chemical and biological agents added to environment by human activities

All issues of environmental health related to water, air, energy and industry are discussed.

WATER-RELATED DISEASES

Water becomes a health hazard on many accounts. Polluted or dirty water and contamination of available water due to poor sanitation and polluted discharges from human habitation and industry harm people.

    ◇   Water-borne or water-washed diseases—e.g. diarrhoea diseases, infectious hepatitis.

    ◇   Water-based diseases—flourosis, dracunculosis.

    ◇   Water-related insect-vector-transmitted diseases—filariasis, malaria and poor personal hygiene due to inadequate quantity of water, e.g. conjunctivitis, scabies.

Poison, Toxic Substances and Health

In popular terms, a poison is a substance that can destroy life rapidly, even when taken in small amounts. Toxic substances are also harmful to health creating lung and heart diseases, kidney damage, skin rashes, cancer, birth defects and health effects.

There are mental depression lowering of intelligence. Obviously, its more toxic a person absorbs or ingests the serious are the effects. The amount of toxic material that enters a person's body is called "the dose". The dose at which the danger starts is called the "threshold". Above the threshold, the hazard increases as the dose increases.

Pesticides on Human Health

◇   Most pesticides are poisonous although the effects and lethal dosages vary from compound to compound.

◇   In 1985, 500,000 bees of insecticides and fungicides were sprayed in US.

◇   Pesticides poisoning include muscle control nervous disorder, loss of mental awareness, eventual death Pesticides induce cancer.

Table 21.1   Waterborne, disease-causing organisms

Name of organism (or) group	Major disease	Primary source
Salmonella typhi	Typhoid fever	Human faeces
Shigella	Bacillary dysentery	Human faeces
Legionella pneumophila and related bacteria	Acute respiratory illness (legionellosis)	Thermally enriched water
Mycobacterium tuberculosis	Tuberculosis	Human respiratory exudates
Enteroviruses		
Polioviruses	Poliomyelitis	Human faeces
Coxsackie viruses A and B	Aseptic meningitis	Human faeces
Echo viruses	Aseptic meningitis	Human faeces
Other enteroviruses	Encephalitis	

(Contd.)

Table 21.1   (Continued)

Name of organism (or) Group	Major disease	Primary source
Hepatitis A virus	Infectious hepatitis	Human faces
Giardia lamblia	Giardiasis	Human and animal faeces
Cryptosporidium	Cryptosporidiosis	Human and animal faeces
Echinococcus	Echinococcosis	Human and animal faeces
Schistosoma	Schistosomiasis	Human and animal faeces
Algae (blue green)		
Anabaena flos-aquae	Gastroenteritis	Natural water
Microcystis aeruginosa	Gastroenteritis	Natural water
Aphanizomenon flos-aquae	Gastroenteritis	Natural water
Schizothrix calciola	Gastroenteritis	Natural water
Virus	Brain fever	Mosquito's which bite pigs and birds
Sunlight, chewing tobacco, smoking, hot drinks, virus, parasites, genetic factor	Cancer	Natural and man made

Food additives, preservations, artificial colours, flavours and sweeteners will have small amount of chemicals that are not natural to the food are poisonous.

Measures to improve health

- Primary health and distribution of drinking water
- Protected water supplies
- Removing rubbish, night soil and cleaning the drains
- Re-organization of medical facilities
- Mother—child care
- Vitamin Zed fund and prevention of epidemics

EFFECTS OF AIR POLLUTION ON HUMAN HEALTH

Children living in highly polluted areas in the United States and Japan have been shown to have a higher-than-normal rate of severe respiratory infections. The pollution of urban air is directly harmful to the health of people with a history of heart disease.

Rain in many parts of the world has become much more strongly acid, due to the effects of air pollution. Much of our acid rainfall has been between pH4 and pH5, but more severely acidic episodes occur from time to time.

Carbon monoxide is a very harmful gas released into air by vehicles and is more concentrated in congested highways to a level of about 100 ppm. Thus, drivers are the most affected people. CO causes difficulty in breathing, headache, and irritation of mucous membranes. It combines with haemoglobin of blood, reducing its O_2-carrying capacity.

At the 10 per cent carboxyhaemoglobin in blood due to smoking there may be lowered tolerance to CO. Cigarette smokers have increased haematocrit (per cent volume of red blood cells) within minutes of smoking. According to some, however, smoking provides immunity to Parkinson's disease affecting nervous system and characterized by tremors, muscular rigidity and emaciation. Pyridine is released into body while smoking and it provides protection against this disease, probably by competing with other toxic substances and blocking the impact on neuroreceptors.

SO_2 causes intense irritation to eyes and respiratory tract. It is absorbed in the upper respiratory tract, leading to swelling and stimulated mucous secretion.

NO_2 causes irritation of alveoli, leading to symptoms resembling emphysema (inflammation) upon prolonged exposure to 1 ppm level. Lung inflammation, may be followed by oedema and finally death.

Mercury inhalation of 1 mg/m^3 of air for three months may lead to death. Nervous system, liver and eyes are damaged. Infants may be deformed. Hydrocarbons have carcinogenic effect on lung.

Prevention and Control of Air Pollution

To check the pollution from vehicular exhaust, exact time of fuel feeding, using gas additives to improve combustion, injecting air in the exhaust to detoxify the toxicants and updating engine designs are advised.

To control evaporation from fuel tank, low-volatile gasoline is used. Gasolines are subjected to slight pressure to prevent the gas from evaporation. Collection of vapours with activated charcoal when the engine is turned off and ignition when the engine is started is yet another method. Use of filters to catch and recycle the escaping gases from the engine helps to control the toxicity of hydrocarbons.

NOISE POLLUTION AND HUMAN HEALTH

Noise has become a permanent part of our lives, due to machinery, industry and technology. Noise harms the body and mind. The main contributors to noise are factories and industries, transportation and religious activity. Horns, sirens, musical instruments, food mixtures, pressure cookers, fan and coolers also play an important role in causing noise pollution. Heavy transport and sound-amplifying systems used by the shopkeepers also cause pollution.

Noise pollution causes disorders in the auditory system and non-auditory system; while the effect on auditory system brings about permanent loss of hearing, in the non-auditory system it affects speech, and causes annoyance, loss of working efficiency and physiological disorders. Some of the examples of physiological disorders are neurosis, anxiety, hypertension, hepatic disease, increase in sweatening, giddiness, nausea and fatigue. Continuous noise causes increase in cholesterol level, heart attack and strokes.

To prevent noise pollution, trees like casuarina, banyan, asoka, neem and tamarind are planted. They drastically absorb reduce sound levels. Other preventive measures are controlling the authorized use of loud-speakers.

ENERGY AND HEALTH

Energy is a prerequisite for socio-economic development and has direct and indirect impacts on health.

Electricity production is a good yardstick of socio-economic development as it is usually accompanied by better infrastructures and health services. In order to strengthen their economies and to provide the economic basis for good health, most developing countries will need to increase their fossil fuel consumption.

People in developed countries use about ten times more commercial energy than those in developing countries and burn approximately 70% of all the fossil fuel used globally. The combustion of fossil fuels, accounting for about 90% of global commercial energy production, is the largest source of greenhouse gases and atmospheric pollution.

Burning of biomass fuels in open fires or inefficient stoves in poorly ventilated houses give off smoke and chemicals that cause respiratory disease with long term cardiovascular effects.

Although there are virtually no direct health effects from the generation of electricity from hydropower, however, the use of electricity gives rise to localized exposures to electromagnetic field that may increase the risk of developing certain cancers. Further, a number of water-related diseases may occur due to the creation of reservoirs and resettlement of populations when dams are built, particularly where water impoundments provide breeding sites for the vectors.

In case of nuclear power plants, there are risks to health for present and future generations from accidents and unsafe disposal of nuclear wastes.

An increase in heatwaves, the geographical extension of vector-borne diseases and the creation of large number of environmental refugees could exert enormous toll on human health, particularly in developing nations.

INDUSTRY AND HEALTH

Industrial practices cause adverse health effects through the release of air and water pollutants and the generation of hazardous wastes. Industrial effluents have polluted many rivers, lakes and coastal environments, especially in developing countries where pollution control is seldom enforced. Furthermore, hazardous wastes are sometimes exported from developed countries to developing countries because the cost of export is lower than the cost of disposal in the country of origin. Usually, there is little concern over the health of the local populations.

Some of the common occupational diseases are silicosis, pneumoconiosis, lead and mercury poisoning and skin diseases. Continued and frequent exposure to noise, especially in industry, give rise to serious health problems.

HIV/AIDS

Acquired Immune Deficiency Syndrome (AIDS) is a disease caused by the human immunodeficiency virus (HIV). The virus attacks the body's immune system making HIV-infected individuals vulnerable to opportunistic infections, cancers and neurological disorders.

The virus which causes AIDS primarily attacks white blood cells called T4 helper cells that are part of the body's internal defence against disease. The virus may also affect the central nervous system.

An infected person's immune system responds by developing antibodies to fight off the virus. The body's ability to produce disease-fighting antibodies eventually becomes limited in HIV-infected persons as the virus reproduces and destroys the body's T4 helper cells.

Even without treatment, some people with HIV infection have no symptoms at all, some have mild health problems, while others have severe health problems associated with AIDS.

HIV

HIV is a lymphocytotropic and neurotropic virus. In an infected person, it is found in almost all the body fluids and organs. It is present in infective dosages in semen, vaginal and cervical secretions and blood. Exchange of these body fluids from HIV-infected individuals can lead to transmission of HIV infection to another person. Semen contains about fifty times higher concentration of the virus as compared to vaginal and cervical secretions and blood. The central nervous system, testes, lymph nodes, etc. act as reservoirs of HIV. The highest concentration of HIV is found, among the body fluids, in the cerebrospinal fluid.

AIDS

Acquired Immune Deficiency Syndrome is a late stage of HIV infection. By the time a diagnosis of AIDS is made, HIV will already have seriously damaged the body's immune system. Often, a person with an AIDS diagnosis will already have had a life-threatening infection or cancer.

Before the use of effective treatment, it commonly took 10 years or more from the time of initial HIV infection to a diagnosis of AIDS; and, on average, it would take another two to four years before death. However, new treatments are radically slowing the destruction of the immune system caused by HIV and lengthening life expectancy. Some people with HIV infection may have never developed AIDS.

HIV in India

Surveillance for HIV infection was initiated in India by the Indian Council of Medical Research (ICMR) in late 1985 as a part of AIDS Task Force. Anti-HIV antibodies were first detected among sex workers from Madras in South India in 1986, and soon thereafter the first AIDS case in India was reported from Bombay in 1986. Realising the high HIV prevalence rates amongst the professional blood donors, anti- HIV screening of all the blood units to be used for transfusional purposes was made mandatory in July 1989 in the four metropolitan cities of India. Presence of HIV infection in India was reported for the first time from Bombay in 1991. Perceiving the challenge posed by 111V epidemic and its potential disastrous consensus, ICMR established the national AIDS Research Institute in Pune in 1992. Two different types of HIV epidemic are seen in India. In North-Eastern India, the epidemic is mainly among intravenous drug users whereas it is mainly spread through sexual route in the rest of the country. Identification of mutations and subtypes are likely to be important in HIV vaccine development.

HIV Epidemic

Predominant route of transmission of HIV is sexual. Hence this epidemic spreads as one would envisage in a classical epidemic of sexually transmitted disease (STD). HIV epidemic spreads classically in three stages. In the first stage, HIV infection is seen amongst sex workers or intravenous drug users (IDUs) which are also called as core transmitters or core groups. In its second stage, HIV infection reaches the clients of sex workers or partners of IDUs. When evidences suggest affection of spouses and children of the clients of sex workers, the HIV epidemic is understood to have reached its third stage. HIV infection is brought into the low-risk population from the core transmitters through the bridge population, a term used to connote mobile population such as truck drivers, single male migrants, etc. Some scientists, consider that IDUs constitute the first stage which is followed by the sex workers and so on. Recently, a fourth stage comprising of adolescents has also been proposed to indicate the severity of the epidemic.

HIV epidemic: global scenario   Though initially the epidemic was mainly reported amongst gay men in the United States and other developed countries, it followed by reports of a similar epidemic hut from the heterosexual population especially from countries in African continent. Within a short span, the epidemic turned into a pandemic. According to the conservative estimates of Joint United Nations Programme on HIV/ AIDS (UNAIDS), by the end of 1998 about 33.4 million people were infected by HIV world over Of these persons, 5.8 million had acquired this infection in 1998 itself. Everyday 16,000 new infections were estimated to be occurring in the world. Of these new infections, 7,000 persons belong to 10–24 years age group. About 40% of the new HIV infections occur amongst women.

More than 95% of the infections are estimated to be occurring in the developing countries. Of those who are HIV-infected, two-thirds are from countries in sub-Saharan Africa. The epidemic was probably introduced late in Asia. However, the rate of rise of HIV infection is alarming in Asia. Of those who are my infected in South and South-East Asia, almost two-thirds are from India. The male to female ratio is 3:4 globally, but it is poised to reach the proportion of 1:1 in sub-Saharan Africa. The number of children infected by HIV is much lesser compared to adults. But the estimate of death due to AIDS reveals a change in the proportions due to their shorter survival with HIV infection when compared to the adults.

However, with the advent of highly active antiretroviral therapy (HAART) , a dramatic decline has been observed in .the death rates due to AIDS in the developed countries.

HIV epidemic: Indian scenario   As a part of the ongoing nationwide surveillance, the National AIDS Control Organization reported that a total of 87,313 persons were found to be HIV positive and 8,220 had developed AIDS by the end of July 1999. These figures are a gross underestimate of the situation due to under-reporting and inconsistent sera-surveillance activities in different states of the country. Estimation reported in 1998 indicate that atleast 0.8% adults between the age group of 15–49 years are infected by HIV in India giving a total of about 4.1 million people. The doubling time of AIDS cases in Vellore, South India was reported to be approximately one year. The doubling time of HIV infection among STD patients in Pune is about two years.

Clinical Manifestation

HIV infection may leads to diseases and illnesses which can take many forms. The problems associated with HIV infection range from the complete absence of symptoms (asymptomatic), to mild illness, to debilitating neurological disorders, to conditions which can lead to death.

The incubation period before any symptoms of HIV disease appear varies significantly from person to person. Some develop symptoms within six months to two years of exposure. Many others, however, may be infected for as many as seven years or more and shown no signs of illness. Research is being done to determine why some infected people become fatally ill while others have milder symptoms or remain symptom-free.

Symptoms of HIV disease include:

◇   loss of appetite

◇   weight loss

◇   tiredness

◇   fever

◇   night sweats

◇   swollen lymph glands

◇   skin rashes

◇   diarrhoea

◇   poor resistance to infection

The condition called AIDS represents a syndrome of late-stage diseases in which the immune system is unable to fight off viruses, bacteria, protozoa, and fungi resulting in infections and diseases which frequently lead to death.

Neurological involvement occurs in many, perhaps a majority of, people with AIDS as the disease progresses. "AIDS dementia (AD)," also called HIV encephalopathy may include auditory, visual, and other sensory disturbances, muscular dysfunction, gross personality changes, memory and judgement impairments and/or a severe loss of intellectual, social or occupational abilities. The most common cause of AD is believed to be direct involvement of the brain by HIV.

Transmission

Unlike flu or measles, HIV is not transmitted through the air; it must get into the bloodstream to cause infection. For this reason, HIV positive people don't pose a risk to others through any form of casual contact. There is no evidence that AIDS is transmitted through coughing, sneezing, food preparation, drinking fountains, toilet seats, mosquitoes, being around an infected person on a daily basis, or donating blood.

HIV is carried in blood, semen, vaginal secretions, breast milk and other body fluids of an infected person. It is transmitted from one person to another by three routes:

1. through sexual intercourse including vaginal intercourse, oral intercourse, and anal intercourse (male to female; female to male; male to male; female to female);
2. through blood-to-blood exposure to infected blood; and
3. from infected women to their infants before, during or shortly after birth.

Major Risk Factors

Persons at increased risk for being infected with HIV include:

◇ present or past drug users
◇ sex partners of drug users
◇ sex partners of infected persons
◇ persons who received blood-clotting factor, blood transfusions or blood products prior to 1986
◇ children born to infected mothers

Prevention

There is no vaccine against AIDS or any treatment so far that can reverse the damage done by HIV to the immune system. People can and must learn how to protect themselves and their loved ones from HIV infection. Following are some basic elements of AIDS information related to prevention. These elements can be adapted to varying degrees of specificity.

Schools and other important community institutions, such as religious organizations, families, and voluntary organizations can adapt the presentation of this information to fit within their value systems while providing scientifically sound, accurate and frank information. Within this framework, individuals should be given information (expert advice, consultation, and crative assistance from health care providers, religious counsellors and health educators) to help them develop responsible life-affirming/health-affirming behaviours which will protect themselves and their contacts from virus transmission.

AIDS Awareness Programme

To create awareness regarding AIDS Red Ribbion clubs are started in many institutions like universities, colleges through NSS activities. Through these clubs awareness programme are conducted regarding AIDS. Red Ribbon indicates love. Love is medicine to those affected by AIDS. These clubs are working for the eradication of fear of people about AIDS.

Figure 21.1   Participants of Red Ribbon Clubs-Awareness program

Many of the people in India affected by AIDS are resource poor and food insecure. The frequent droughts in various states and districts over the years have had a serious impact on the food security and nutritional status of the large population living in these areas.

One way to mitigate the impact of HIV/AIDS is to strengthen food safety nets for infected and affected people. Thus food and nutrition intervention can play a vital role in care, treatment and mitigation of the effects of HIV/AIDS.

WOMEN AND CHILD WELFARE

In developed countries maternal death rate is between 2–140 per 100,000 live births, whereas in the developing nations this rate is between 8–1100 per 100,000 live births. By improving the health of mothers and children, we contribute to the health of the general population. The main causes are lack of proper education, malnutrition, infections and unregulated fertility.

Pregnant women, nursing mothers and children are particularly vulnerable to the effects of malnutrition. The adverse effects of malnutrition of women are—maternal depletion, low birth weight, anaemia, toxemias of pregnancy, post-partum haemorrhage, all leading to high mortality and morbidity.

Measures to Improve the Nutritional Status of Women and Children

Distribution of iron and folic acid tablets, fortification and enrichment of foods, nutrition education, etc. Indirect interventions have still wider ramifications, as they are not specifically related to nutrition. Clean drinking water, family planning, food hygiene, education and primary health care use some other measures to improve the nutritional status.

Infections

Women or maternal infections can cause a variety of adverse effects such as threatened abortions, foetal growth retardation, low birth weight, embryopathy and puerperal sepsis. Also, about 25% of women in rural areas suffer at least one out of urinary infections.

Children may be ill with debilitating diarrhoeal, respiratory and skin infections, or the situation is further aggravated by chronic infections, such as malaria and tuberculosis.

WHO, expanded programme on immunization. Developing countries are being immunized against tuberculosis, diphtheria, pertussis (whooping cough), tetanus, measles and polio. Women and children are educated on personal hygiene and on sanitation measures to eradicate common parasitic diseases and infections.

Development of Women and Child/Welfare

Under the scheme of Balika Samridhi Yojana launched on October 2, 1997, with a specific sole objective to encourage the enrolment and retention of a girl child in the schools, the mother of a girl child born on or after August 15, 1997 in a family below poverty line is given a grant of Rs. 500 besides a scholarship for education of the girl child when she attends school. About 12 lakh girl children were benefited during 1997–98: Under Indira Mahila Yojan a for empowerment of women, 28,000 small homogeneous women's groups were formed up to 1997–99. The scheme of Mahila Samridhi Yojana to inculcate the habit of saving among rural women is being revised and merged into Indira Mahila Yojana to have an integrated package of five components including formation of viable women's groups.

Child deveopment   Child development and welfare centres are day care centres, Balwadi Nutrition Programme, Early Childhood Education and National Institute of Public Cooperation and Child Development. In Articles 39 (f) of Constitution of India protects towards securing childhood and youth against moral and material abandonment.

ROLE OF INFORMATION TECHNOLOGY (IT) IN ENVIRONMENT AND HUMAN HEALTH

IT is a tool that can be effectively used in planning and organizing diverse activities such as eradication of poverty, power generation, administrative reforms, etc. Today IT can be used in bringing the government and people to work together in realizing the goals whether it is education, ecology, health or any other sector. IT is useful for improving the quality of health care at the individual medical practitioner's level. IT is useful for medical education. IT can be a very effective and helpful tool for the doctor in managing patient records.

IT is useful in information storage media and information transfer mechanisms and referred to the new technologies of remote sensing, Geographical Information System (GIS) and the availability of a number of databases for better environment management. The Ministry of Environment of the Government of India has taken efforts in developing and maintaining ENVIS (the environmental information system), and the databases developed by the Indira Gandhi Conservation Monitoring Centre. IT is useful in the management of micro-level environmental issues such as industrial pollution.

◇ IT is a tool in local environmental issues in our eagerness to be global.

◇ There is a discipline of "Environmental Informatics" with the primary objective of using IT for assistance in monitoring, control, information management, computation and analysis, and decision support in environmental management.

◇ The major challenges facing the country through IT are population control, conservation of non-renewable energy sources, conservation of forests, re-orientation in user and management of resources, developing appropriate science and technology strategies and tools in the production of goods and services to limit environmental pollution. There is a need for development of adequate information infrastructure and application of IT in this regard.

◇ Steps are to be taken to boost IT for agricultural and integrated rural development in the country.

◇ It is recommended to have a national policy on information security, privacy and data protection for handling computerized data.

Conservation of Bio-resources

Bioiverity conservation   Use of GIS (Geographic Information System) and Remote Sensing can help in determining the rates, causes and scale of biodiversity loss. Information on deforestation. and land use change can be integrated with data on the distribution of biodiversity and existing information on climate, topography, soil, etc.

Species monitoring   Not only the existence of flora and faura can be detected, even counting of animals like elephants, tigers, etc. can be done with the help of GIS. Information on endangered and vulnerable species of plants and animals is kept in a database which acts as an aid in appreciating the degree of danger that a species is in.

Plaanning seection   Railway routes, dam or reservoir sites, waste, disposal sites, major industrial sites, etc. can be planned in such a way that they cause minimal disturbance to ecosystems.

Disaster management   Data can be utilized for obtaining quantitative estimates of the damage to infrastructure and property caused by earthquakes, cyclones, floods, landslides and volcanic eruptions. Disaster-prone areas can be identified where appropriate action can be taken up to reduce the losses and also disasters like cyclones, floods, etc. can be predicted well in advance.

IT and Health

Information technology can lead to meaningful action for improving the health of the society in such vital areas as population control, environmental protection, prevention of accidents, nutrition care, etc.

IT is helping in the development and application of computational tools to acquire, store, organize, archive, analyse and visualize satellite and biological data.

Information technology which is applied in the field of Biology is named as Bioinformatics. Bioinformatics is useful for developing databases of bioresources and of traditional knowledge in medicine.

The bioinformatics study began with the analysis of osteoclasts. In normal course it would have taken months, if not years, to find the cause of the disease at the molecular level. But bioinformatics is now user-friendly and provides tools to analyse, store and organize data. The tools of IT help in studying the huge biological data analysis and the associated problems to plan further experiments. Bioinformatics played a key role in the final stages of the Human Genome Project. The race was on and the competition to reach the goal was intense.

DNA database is databanks having genetic information. Many organizations, such as WHO, maintain their websites with information about endemic, epidemic and communicable diseases. New drug release, their mode of action and the field of surgery is also available. Telemedicine and distance medicine is now far-reaching along with documentation and display of human anatomy with the help of internet. The use of bioinformatics is not limited to the above discussed areas of human health, but is used in various other fields including taxonomic and phylogenetic studies.

The subject of health is as important as agriculture and industry. Panchayati Raj be entrusted with the jobs to provide primary health care and information technology be used to monitor the working of Panchayats.

The Panchayati Raj should provide the necessary condition for people's health sector where a bottom-up rather than top-down approach to solve community's health and medical problems.

An agency such as the Institute of Environmental Health Sciences (IEHS) be established on priority. The institute can play a major role in planning of resources (manpower and funding) required to meet the current and future needs of the public environmental health concerns. IT can also serves as a national resource for environmental health information, training and education and environmental awareness among the health, R&D professionals including the public at large.

IT is useful in environmental cleanliness and sanitation so as to reduce the infections in children. Education of the public through media and public forum about the preventive measures and need for strengthening government run secondary and tertiary care medical hospitals and institutes.

Disease profile is going to be influenced by life-style, special efforts need to be made to monitor, evaluate and control these emerging maladies through IT.

HUMAN RIGHTS IN INDIA

Human rights means, the rights relating to life, liberty, equality and dignity of the individual guaranteed by the constitution or embodied in the International covenant on economic, social and cultural rights adopted on December 6, 1996 and enforceable by courts in India. A "National Human Rights Commission (NHRC)" was set up by India in October 1993. The Commission comprises five members and a chairman to head. Continuous attempts are being made by the commission to address various human rights issues. Some of these issues are being monitored as programmes on the directions of the supreme court.

The programmes in pursuance are

◇   Abolition of Bonded Labour

◇   Functioning of the Mental Hospitals at Ranchi, Agra and Gwalior

◇   Functioning of the Government Protective Home (Women), Agra

◇   Right to Food

Other Programmes and Human Rights issues taken up by the Commission include

- ◇ Review of the Child Marriage Restraint Act, 1929
- ◇ Protocols to the Convention on the Rights of the Child
- ◇ Preventing Employment of Children
- ◇ Abolition of Child Labour
- ◇ Trafficking in Women and Children: Manual for the Judiciary for Gender Sensitization.
- ◇ Sensitization Programme on Prevention of Sex Tourism and Trafficking
- ◇ Maternal Anaemia and Human Rights
- ◇ Rehabilitation of Destitute Women in Vrindavan
- ◇ Combating Sexual Harassment of Women at the Work Place Harassment of Women Passengers in Trains
- ◇ Abolition of Manual Scavenging
- ◇ Dalits issues including atrocities perpetrated on them
- ◇ Problems faced by Denotified and Nomadic Tribes
- ◇ Rights of the Disabled
- ◇ Right to Health
- ◇ HIV/AIDS
- ◇ Relief Work for the Victims of 1999 Orissa Cyclone
- ◇ Monitoring of relief measures undertaken after Gujarat Earthquake (2001)
- ◇ District Complaints Authority
- ◇ Population Policy—Development and Human Rights

Functions of the Commission

- ◇ To inquire, on its own initiative or on a petition presented to it by a victim or any person on his behalf, into complaints of violation of human rights or abetment thereof or negligence in the prevention of such violation by a public servant.

- ◇ Also the commission can review the safeguards provided by our constitution or law for the protection of human rights and recommend measures for their effective implementation.

◇　While inquiring into complaints under this Act, the commission shall have all the powers of a civil court trying a suit under the Civil Procedure Code, 1908.

Powers of the Commission

◇　To summon and enforce the attendance of witness and examine them on oath;

◇　To discover and produce any document;

◇　To receive evidence on affidavit;

◇　To requisition of any public record or copy there of from any court of office; and

◇　To issue commissions for the examination of witnesses or documents;

◇　The commission, besides its own investigating staff, can also utilize the agency of the central government or any state government as and when required;

◇　It has also taken up the task of protection of human rights in areas of insurgency and terrorism;

◇　It is making all possible efforts to prevent custodial deaths, tortures and rape;

◇　Commission also reviews laws, implementation of treaties and other instruments of human rights literacy, and encourages the efforts of nongovernmental organisations and institutions working in the field of human rights.

Case Study 20

Abolition of Child Labour

The NHRC has been deeply concerned about the employment of child labour in the country as it leads to denial of the basic human rights of children guaranteed by the Constitution and the International covenants.

The commission on child labour has observed that "No economic or social issue has been of such compelling concern to the commission as the persistence, fifty years after Independence, of widespread child labour in our country. It prevails, despite articles 23, 24, 39 (e) and (f), 41, 45 and 47 of the constitution and despite the passing of various legislations on the subject between 1948 and 1986. It has defined the terms of six conventions of the International Labour Organization to which India is a party and the Convention on the Rights of the Child, in addition. Despite the announcement of a National Child Labour Policy in 1987, the subsequent constitution of a National Authority for the Elimination of Child Labour (NAECL) and the undertaking of National Child Labour Projects (NCLP) in an increasing number of areas of our country, the goal of ending child labour remains elusive, even in respect of the estimated two million children working in hazardous industries who were to be free from such tyranny by the year 2000".

The commission is focusing its attention on the following industries where from rampant reports of child labour were received. These interalia include the

- Bangle/glass industry
- Silk industry
- Lock industry
- Stone-quarries
- Brick kiln
- Diamond cutting
- Ship-breaking
- Construction work
- Carpet-weaving

The commission monitors the child labour situation in the country through its special rapporteurs, visits by members, sensitization programmes and workshops, launching projects, interaction with the industry associations and other concerned agencies, coordination with the State Governments and NGOs to ensure that adequate steps are taken to eradicate child labour.

The commission believes that unless and until the reality of free and compulsory education for all up to the age of 14 years is realized, the problem of child labour shall continue. The commission has involved the NGO sector in the non-formal education of child labourers and a number of such schools/training centres are functioning in the districts of the carpet belt. There has also been a distinct improvement in the level of awareness among the general public about child labour issues.

Case Study 21

Human Rights - Global

Declaration of Human Rights 1948.

In 1948, the United Nations has adopted and proclaimed the Declaration of human rights. The declaration states that all members of the human family is the foundation of freedom, justice and peace in the world.

The fundamental rights in this declaration are :

- the right to life, liberty and security of persons
- the right to own property
- the right to an adequate standard of living
- the right to education, freedom and religion
- the right to freedom

Examples of some sections of Article of declaration

Article 3   : right to life, liberty and security of human being

Article 10 : determination of human rights and obligation

Article 15 : right to nationality

Article 19 : right to freedom of operation

Article 26 : right to education

VALUE EDUCATION

Value education is essentially for personality and character buildings and moral cultural and spiritual development. Value education is as important as formal education because if a highly qualified, well-employed person does not know how to behave properly, then all that he does has little meaning and will not serve him well.

Value is to do not only with beliefs but also with our understanding, feelings and behaviour. Value education may be understood in a broad sense to mean all aspects of the process by which teachers (and other adults) transmit values to pupils.

The values most often spoken of should be

- consideration for others
- respect for others, and for property and authority

- religious values, a religious community
- work
- self-esteem
- self-discipline

Teaching Moral and Value Education

Value education needs teaching tolerance. Tolerance means accepting other and appreciating differences. In addition, tolerance means many other behaviours and attitudes. It means

- being able to control one's temper and anger,
- being patient,
- being able to live under stress,
- being able to cope with hardships,
- being able to accommodate different points of view,
- being able to forgive and forget.

So, as we see it means much more than only accepting others and their differences, it deals a great deal with inner feelings and behaviours.

Value education can be taught through the two-way approach strategy which includes

1. Integrating the concept in the content of all subjects across the curriculum
2. Teaching moral and value education

Integrating the concept in the content of all subjects across the currivulum Some of the issues to be taught are:

- Human rights
- Gender and preventing discrimination practices against women
- Democracy
- Child rights
- Tolerance
- Education for peace
- Tourism

- ◇ Protective and curative health
- ◇ Education for citizenship
- ◇ Environment, how to protect it and keep it beautiful
- ◇ Overpopulation and development
- ◇ Consumer education
- ◇ National unity.
- ◇ Eliminating all types of fanaticism.

How to teach

1. Analyse every issue into minor and major concepts, which are then designed as a conceptual map.
2. Concepts were then categorized and sequenced into a matrix.
3. Experts choose the relevant concepts to their area of specialization, and the level of the concepts suitable to school or college level.
4. As such concepts are integrated in all the textbooks and we also emphasize methods of handling those concepts in the teachersguides.
5. Teachers who teach different subjects attend training programmes and workshops to familiarize themselves with these concepts and issues, and how to teach them in the classroom.
6. The activity book which accompanies the text book also includes activities and questions related to those issues.
7. Special teaching/learning enriching materials are developed to foster the learning of those concepts.
8. Use of special educational kits, multimedia computer programmes, and television video tapes.

Teaching moral and value education   Under this moral and value education students learn:

- ◇ to love one another.
- ◇ to work together.
- ◇ to appreciate friendship.
- ◇ how to be responsible.

- ⬦   to be fair and kind.
- ⬦   to be honest in what they say and do.
- ⬦   the importance of freedom to everyone.
- ⬦   the relation between one's rights and others.
- ⬦   the relation between one's rights and duties.
- ⬦   how to respect oneself and the others.
- ⬦   to understand the importance of being broad-minded and that there are more than one side to everything.
- ⬦   that life depends on cooperation, and together we are stronger than any one alone.
- ⬦   to discuss issues to solve problems.
- ⬦   to negotiate.
- ⬦   to appreciate living in peace.
- ⬦   to enjoy happiness.
- ⬦   to avoid violence.
- ⬦   to know that every human being has worth and dignity.
- ⬦   to avoid lying.
- ⬦   to admire courage.
- ⬦   to enjoy success and feel happy for the success of others...

How to teach is more important than what to teach. Therefore, the course promote active learning, be student centered and have everyday life related content. It is better that value education be presented in a story format and as real daily life situations which require students to think, reflect and discuss. Thus they infer the value in the story.

The acceptable system is the one in which we learn to share our exhaustible resources—to regain a balance. This requires that we reduce our needs and that the materials we use must be replenishable.We must treat all of the earth as a sacred trust to be used so that its contents are neither diminished nor permanently changed. The recognition of the need for such adaptation, called environmental ethics, has developed into what we now call environmental ethics.

"Environmental Ethics" is a set of principles and rules to govern human behaviour. Such an ethic can be viewed as an adaptive device, which should enable the human species to thrive in the world. It is destined not only for individuals, but also for communities, governments, non-governmental organizations, businesses, etc. In order to be effective, it should be understood by and acceptable to peoples of all races, religions, philosophies, etc.

The facts, which are being brought to light about what is happening in the world around us, are fraught with terrifying consequences. The damage to or depletion of the ozone layer, which protects the earth from much injurious cosmic radiation, the increase in global temperature, which might result in the melting of the ice caps and a calamitous rise of the sea level, and the running out of resources essential for the continuance of civilized life as we know it now are all daunting.

CONSUMERISM AND WASTE PRODUCTS

Consumerism is the chronic purchasing of new goods and services, with little attention to their true needs, durability, product origin or the environmental consequences of manufacture and disposal. Consumerism is driven by huge sums spent on advertising. This creates impact in environment by consuming resources and disposal of waste.

By the 1930s, the world economy reached a point when the capitalist class started to manufacture needs and desires and to raise the level of consumption so that this class could continue accumulating capital; the system had reached a point where the appetites of capitalists exceeded the demand of consumers for basic goods and services. So began the impetus for, and the creation of mass consumerism.

By the 1950s, a whole industry led by public relations and marketing had developed, which functioned to help the capitalist class grab the surplus value from high production levels under "consumerism." Massive consumerism was in force, and some troubling consequences were starting to be felt.

Several important tactics that capitalists use to generate consumerism are the following.

- ◇ Creating new psychological "needs" in people
- ◇ Stimulating impulse buying
- ◇ Creating and marketing fads and styles to spur temporary "usefulness" of material goods (or social obsolescence)
- ◇ Making short-lived or hard-to-fix goods; many products are designed to have uneconomically short lives, with the intention of forcing consumers to repurchase too frequently. This is technical obsolescence, often called "planned obsolescence".

Effects of Consumerism

- ◇ Increased consumption is related to pollution—consumption itself and the production processes generate waste products (e.g. air pollution by automobiles, industrial waste waters).
- ◇ 86 per cent of the world's resources are consumed by the rich who constitute only 20 per cent of the world's population.
- ◇ Even through pollution is occurring in poor countries, a large portion of it is to meet the consumer demand of the developed western countries. (e.g. all leather goods are manufactured in Indian cites like Vellore, Vaniambadi and exported to Germany).

⋄   Most of e-waste is dumped in developing countries like Asian countries.

Overcoming Consumerism

The time has come for consumers to take the lead in prompting manufacturers to adopt clean and eco-friendly technologies and environmentally safe disposal of used products, along with preventive and mitigative approaches.

The issues of environmental protection has brought the consumers, the industry, and the government to a common platform where each has to play its own role. Some of the fields that have already been well-developed include industrial ecology, green marketing, green consumerism, pigouvian taxes and eco-labelling.

Industrial ecology   Application of the ecosystem concept to industries—linking the metabolism of one comparing with other, e.g. sewage, agri waste and household refuse disposal are like of with district electricity generation in Germany and Sweden.

Green market   Marketing improved products to recognize.

Green consumerism   Along the growth in green marketing, there has been green consumerism. There is consumer production bodies since 1960s.

Piguouvian tax   Making manufacturers representative for some or all of the costs of recycling or waste disposal. German companies pay a fee to display green dot (Ecomark (Figure 22.1)) on packing which authorizes recycling. As the costs have to be borne by customers, this encourages companies to reduce expensive packing.

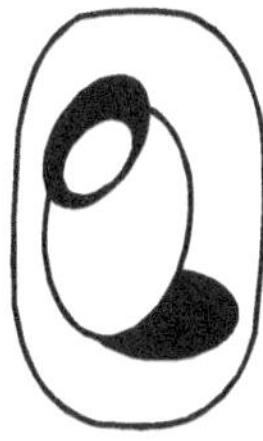

Figure 22.1   EcoMark India (1991)

One can lead a good life through elimination of wastes and using products by sustainable means. Waste often is found by discovering patterns in how we live, our choice of what to buy, what to consume and what to do with the by-products.

Simple Ways to Overcome Waste

These are simple everyday ways to overcome consumerism.

- Every item that passes through one's hands should be utilized to its maximum.
- Items are useful if they have any utility or recyclable materials left in them that can be used.
- Every serviceable item that one discards should be treated as potential resource that can be used somehow.
- Recycling and reuse reduces wastes.
- Making an old item to be used for a different purpose is an intellectual challenge and can be far more satisfying than spending money for a new item.
- Education is the best way to overcome consumerism.

NEED FOR ENVIRONMENTAL ETHICS

Nowadays environmental problems are increasing in magnitude. There is seriousness of environmental problems as there is increasing power and diversity of technology on one hand and, an exponential growth of the human population, which is unequally distributed over the earth's surface on the other.

Now environmental problems are global, they are also very much local and present at our very doorsteps. Dirt and refusals thrown about, the effluents from factories and industrial units, large and small, the ill effects of the indiscriminate use of pesticides, etc., are being experienced almost everywhere.

Humans are now able to affect the biosphere on a scale comparable to that of natural geophysical and geochemical agents and its processes. Because of human activity there is loss of biological species, which are irreversible and yet, human life is submitted to the same ecological processes as the other parts of the biosphere.

It has become apparent that technological development alone is unlikely to enable humans to solve the generalized problems. The changes in attitude and behaviour are needed to modify the patterns of resource depletion, pollution, and waste accumulation or, in other words, to strive for sustainable development in industrialized as well as in developing countries. This should reconcile high-quality economic growth with the preservation of the environment.

In addition to external sanctions for doing what is needed, internal sanctions are also required, i.e., the adoption of an ethic. People living today not only have to consider their own well-being, but also the well-being of future generations. Also we must remember that we are dependent on the natural environment not only for our physical needs but also for our psychic needs as well. A large number of plants and animals have enormous aesthetic values, medicinal values and economic value. Hence all of us have an obligation to safeguard and to protect our environment.

Environmental ethics shows us the importance of respecting, appreciating, protecting and regenerating environment. It evokes a love for nature. It forces us to use the gifts of nature judiciously avoiding waste and destruction. It forbids us to abuse natural resources. It helps us to create awareness in the minds of all people. It makes us to prevent pollution, deforestation and depletion of natural resources. It prompts us to act to form nature clubs and to motivate people to work for a clean and healthy environment.

CODE OF ENVIRONMENTAL ETHICS

Once an ethical foundation is set, codes of environmental practice can be developed for individuals, communities, governments, businesses, and so on, e.g. the codes advocated by various business groups or non-governmental organizations such as the International Chamber of Commerce or the Keidanren (Business Association of Japan). Some international agreements include reference to ethical principles. For example, the UNCED Convention on Biological Diversity is based on the remarkable acceptance of the intrinsic value of biological diversity.

The proposed codes are tailored to apply ethical principles to their potential users, taking into account the state of scientific knowledge, the nature of the problems addressed, the socio-economic circumstances, the legal and regulatory backgrounds, cultural situations, etc.

A code of environmental practice of general application, which has been developed by an international group of people following the 1989 Economic Summit Nations Conference on Bioethics (Berry, 1993). It is established on the ethic of stewardship of the living and non-living systems of the earth, in order to maintain their sustainability for the present and the future, allowing development with equity. This ethic is based on the following premises.

⋄   The need for scientific knowledge on ecological processes and the state of the environment, in order to inform the political decision-making process;

⋄   The full accounting of external and internal costs;

⋄   The recognition of the interdependence of environmental values, such as the demands of present and future generations and of nature itself.

A number of guidelines are formulated on the basis of the above-mentioned principles, which are as follows:

1.   Safeguard the ecological processes essential for the proper functioning of the biosphere;

2.   Integrate natural resource conservation with socio-economic development;

3.   Improve the understanding of the environment through research, education, access, and dissemination of information;

4.   Protection of global commons such as atmosphere, deep-sea, regions covered by the Antarctic Treaty System, etc. through international cooperation;

5.   Avoid irreversible choices and apply the precautionary principles.

Practical Code of Environmental Ethics

Detailed obligations have been derived to apply these guidelines to individuals, communities, governments, businesses, etc. They are as follows:

⋄   All environmental impacts should be fully assessed in advance, for their effects on the community, posterity, and nature itself, as well as on individual interests.

◇   Regular monitoring of the state of the environment should be undertaken and the data must be made available without restriction.

◇   The accounting of activities involving environmental impacts should incorporate social, cultural and environmental costs, as well as commercial considerations.

◇   The facilitation of technological transfer should be guaranteed with justice to those who develop new technologies and equitable compassion towards those who need them.

◇   Regulatory and mandatory restrictions should be controlled by cooperation rather than confrontation.

◇   Minimum environmental standards must be effectively monitored and enforced, and a regular review of environmental standards and practices should be undertaken by expert independent bodies.

◇   Costs of environmental damage should be fully borne by their instigator.

◇   Existing and future international conventions dealing with trans-boundary pollution or the management of shared natural resources should include

1.   the assumption of the responsibility of every state not to harm the health and environment of other nations;

2.   liability and compensation for any damage caused by third parties; and

3.   equal right of access to remedial measures by all parties concerned.

◇   Industrial and domestic wastes should be reduced as much as possible, if appropriate, by taxation and penalties on refuse dumping. Waste transport should be minimized by adequate provision of recycling and treatment plants.

◇   Appropriate sanctions should be imposed on the selling or export of technology or equipment that fails to meet the best practicable environmental option for any situation.

◇   International agreement should be sought on the management of extranational resources (global commons).

ENVIRONMENTAL LAWS

In order to achieve a sustainable and clean environment, environmental laws are necessary. The quality of the environment is sustainably declining due to the increasing pollution, loss of vegetation and biological diversity, excessive concentration of harmful chemicals in the ambient atmosphere and in food chains, growing risks of environmental accidents and threats to life-support systems. It is protected by protecting available resources such as forests, lakes, rivers, wildlife and air, etc.

The Parliament of India has passed bills to enforce laws. The main aim of the laws include the following.

- ◇ Regulation of resources
- ◇ Protecting the environment and biodiversity
- ◇ Mediating resolving conflicts
- ◇ Formulating stable, understanding agreements

Some countries have been active in developing environmental laws notably Sweden and the Netherlands. United Kingdom has environmental laws as early as AD 12th century.

The laws in action are given in the following sections.

ENVIRONMENTAL PROTECTION ACT, 1986

After the 1972 United Nations Conference on the Human Environment at Stockholm, India enhanced the Environmental Protection Act, 1986, and the corresponding environmental (protection) rules in 1986. The Central

Government has been given powers to prevent, control and abate environmental pollution.

The various provisions of this act have been amended from time to time.

Salient Features

◇   Protection and improvement of the environment

◇   Prevention of hazards to all living creatures

◇   Maintaining relationship between human and their environment

The Central Government should take the steps:

> ◇   to coordinate action of state government officers and other authorities.
>
> ◇   to lay down standards for the quality of environment.
>
> ◇   to prevent accidents affecting the environment.
>
> ◇   to protect the environment and control environmental pollution.
>
> ◇   to check manufacturing processes materials, and substances likely to fight against air pollution.
>
> ◇   to carry research on environmental pollution problems.
>
> ◇   to take steps to implement the provisions of the act.
>
> ◇   to constitute an authority, for the purpose of performing such functions and powers of Central Government.

Powers of the Central Government Under This Act

Power of entry and inspection   According to Section 10, the Central Government officers have the power to enter and inspect any place for the purpose of performing any function entrusted under the legislation.

Any person carrying on an industrial, operation or process or handling any hazardous substance shall have to render all assistance to the Central Government and its officers failing or obstructing which will be a punishable offence under this Act.

Power to take samples   Under Section 11, Central Government and its officers have the power to take samples of air, water, soil or substances from the factory or place for analysis. In case the occupier or the agent refuses to cooperate, the refusal can be reported in writing to the government analyst.

Power to establish laboratories   Under Section 12, the Central Government has the power to establish environmental laboratories or recognize any laboratory or institute as an environmental laboratory.

Under Section 13, the Central Government has the power to appoint or recognize government analysts for the purpose of analysis of samples of air, water, soil or any other substance.

The Central Government has the power to close, prohibit or regulate any industry or operation or to stop or regulate the supply of electricity or water or any other service.

Provisions

Section 7.1   No person in industry shall be permitted to discharge wastewater in excess of prescribed standard.

Section 8.2   No person shall handle hazardous substances without complying with the prescribed procedural safeguards.

Section 9.3   A person is responsible to mitigate environmental pollution and to intimate any occurrences relating to environmental pollution to the concerned authorities.

Section 10   The central government and its officers have the power to enter and inspect any place for the purpose of performing any function entrusted under the legislation.

Section 14   The report signed by a government analyst may be used as an evidence of the facts stated therein in any proceeding under the legislation.

Any person violating any of the provisions of the act shall be punishable with an imprisonment for a term—which may extend up to five years—or with a fine—which may extend to one lakh rupees—or with both.

In case the violation or contravention continues beyond a period of one year after the date of conviction, the offender will be punished with imprisonment for a term of seven years.

According to Section 16, in the case of an offence being committed by a company, the criminal liability is fixed on the company's directors and principal officers.

Under Section 17, criminal liability is also fixed on the heads of department of government where an offence is committed by the concerned department and the head of the department is unable to prove that the offence was committed without his knowledge or that he exercised all diligence to prevent the commission of such an offence. The main concern of the act is to maintain or restore the wholesome of water. Any pollution of physical, chemical, biological changes in excess in water because of discharge of sewage trade effluents will be harmful to public health, to domestic, commercial, industrial and agricultural uses, and to life and health of animals plants or aquatic organisms.

According to this act, the Central Pollution Control Board (CPCB) and State Pollution Control Board (SPCB) have been constituted

Functions of CPCB

1. To promote cleanliness of streams and wells in different areas of the state.

2. To advise and support the Central Government on matters related to prevention and control of water pollution.

3. To coordinate the actions of the state board and to settle the problems among them.

4. To provide technical support and guidance to the state boards to promote research and development work in prevention and control of water pollution problems.

5. To organize seminars and training programmes for those involved in pollution control.

6. To conduct comprehensive programmes for pollution control through mass media—T.V, radio, newspaper and magazines.

7. To prescribe and set standards for streams or wells.

8. To prepare instruction manuals, information brochures, codes or guides for the treatment and disposal of sewage and industrial effluents.

9. To set up and recognize laboratories for testing and analysis of water samples from any stream, well or trade effluents.

Functions of SPCB

According to Section 7-B, the following are the functions of SPCB.

1. Planning a comprehensive programme for prevention, control and abatement of pollution of streams and wells.

2. Advising the State Govt. on matters related to water pollution control or location of industries.

3. Encouraging the R&D work and conducting investigations related to different aspects of water pollution.

4. Cooperating with central pollution control board for training personnel to handle water pollution programmes and organizing related mass education programmes.

5. Inspecting industrial effluents and sewage treatment plants.

6. Setting effluent standards for sewage and industrial effluents.

7. Developing economical and reliable methods of disposal, treatment and reuse of wastewater in agriculture.

8. Setting standards for treatment of sewage and industrial effluents to be discharged into any water bodies.

9. Formulating vary or revoke any order for control of discharge of waste into streams and wells or for construction of systems for disposal of effluents.

10. To set up and recognize laboratories for testing and analysis of samples.

11. To perform functions related to pollution control entrusted by CPCB or SPCB.

Powers of State Government

◇ It has the power to collect samples of water of any stream or well or any effluent.

◇ SPCB has the power to obtain a report of the results of the analysis from any laboratory.

Under Section 23.5, the SPCB is empowered by the state government to enter any place for the purpose of performing any of the functions entrusted to it.

No person shall knowingly allow the entry of any matter into any stream, which may impede the proper flow of water resulting in substantial aggravation of water pollution.

No person shall knowingly allow the entry of any toxic, noxious or polluting matter into any stream, well or sewer or land either directly or indirectly. If a person failed to provide information about discharging effluents into streams or well or regarding construction or establishment of an effluent disposal system, the penalty is imprisonment up to three months or fine up to Rs. 10,000 or both. If the omission continues, the offender is penalized with an additional fine up to Rs. 5, 000 per day.

In the case of violation of order prohibiting discharge of any pollution matter into stream, well or land, or violation of the board's order of closure of industry or stoppage of electricity or water supply, etc. the penalty is imprisonment for one-and a-half years to six years or fine or both.

AIR ACT

It is interesting to note that the control of water pollution was India's own initiative,' while the measures on air pollution came, only as a response to the 1972 Stockholm Conference. The United Nations Conference on the Human Environment at Stockholm in 1972 asked all the participating states to take appropriate steps for the preservation of natural resources of the earth, including the preservation of the quality of air and control of air pollution. The salient features of this act are briefly described below.

Objectives of the Act

- ◊ Prevention, control and abatement of air pollution
- ◊ Maintaining the quality of air
- ◊ Establishment of Boards for the prevention and control of air pollution

The functions of CPCB and SPCB for the control of air pollution are also identical with that of the water pollution except that they aim to control air pollution instead of water pollution.

WILDLIFE ACT

Objectives of the Act

According to Section 1 of this Act, the objectives of this act are

  ◇   to maintain essential ecological processes and life-supporting systems,

  ◇   to preserve biodiversity and

  ◇   to ensure a continuous use of species, i.e., protection and conservation of wildlife.

Contents of Wildlife Act

1. Wildlife includes any animal—bee, butterfly, crustacean, fish, moth and aquatic and land vegetation—which forms part of any habitat.

2. Habitat includes land, water or vegetation which is the natural home of any wild animal.

3. Hunting—to capture, kill, poison, share and trap any wild animal or trying to do so. To injure, destroy or take away any part of the body of such animal and damaging or disturbing the eggs or nests of wild birds and reptiles.

4. Animal articles—according to the act include any article made from any part of a captive or wild animal.

Chief Wildlife Warden

Under Section 3, the Central Government can employ Chief Wildlife Warden and other members of Wildlife Advisory Board. They will declare sanctuaries, National Park, game reserves and closed areas.

Under Section 45, the Chief Wildlife Warden or Authorized Officer may suspend or cancel any licence granted under Section 44, after recording the reasons in writing and giving reasonable hearing to the license holder.

Sanctuary   Under Section 18 of the Act, if the State Government considers any area to be of adequate ecological, floral, faunal, geomorphological, natural or zoological significance, then it may declare such an area as a sanctuary by notification for the purpose of protecting, propagating or developing wildlife therein or its environment.

National park   Under Section 35, if the State Government feels any area is of adequate ecological, floral, faunal, geomorphological, or zoological importance, then it may, by notification, declare such an area as National Park for the purpose of protecting, propagating or developing wildlife therein of its environment.

Closed area   Under Section 37, the State Government may, by notification, declare any area as closed to hunting for a specified period. No hunting of any wild animal shall be permitted in such a closed area.

Zoo authority   Under Section 38-A, the Central Government shall constitute the Central Zoo Authority. They decide the following

1. Specification of the minimum standards for housing, upkeep and veterinary care of the animals kept in a zoo.
2. Evaluation and assessment of the functioning of the zoos.
3. Recognize and de-recognize zoos.
4. Identification of endangered species for the purpose of captive (*ex situ*) breeding.
5. Coordination of acquisition, exchange and loading of animals for breeding purposes.

Hunting   Under Section 9, no person shall hunt any wild animal except as provided under Section 11 and 12 of this Act.

- ⋄ If the Chief Wildlife Warden is satisfied that any wild animal has become dangerous to human life or is disabled or diseased beyond recovery, then he may, by an order in writing and stating the reasons thereof, permit any person to hunt such an animal.
- ⋄ The killing or wounding of any wild animal in good faith, i.e., in self-defence or defence of any other person, shall not be an offence.
- ⋄ Any wild animal killed or wounded in defence of any person shall be the government property.

Animal articles   According to Section 44 of this Act, no person shall, except under and in accordance with a licence, carry out business as

i. Manufacturer of dealer in animal articles.
ii. Dealer in "trophy" (rugs, skins, specimens of animals mounted in whole or in part antler, horn, hair, feathers, musk, tooth, nest, eggs) or captive animal or meat.

iii.   Taxidermist (curing, preparation or preservation of trophies).

iv.   Cook or serve meat in any eatery.

Penalties

1.   A person violating any of the provisions of this Act shall be punishable with imprisonment for three years or with a fine of Rs. 23,000 or with both.

2.   In case a person is convicted of an offence against this Act, the court may order that any captive animal, wild animal, animal article, trophy, weapon, vehicle, trap, etc. be forfeited to the state government, and that any permit or license held by such a person be cancelled in addition to the other penalties awarded in such an offence.

3.   In case of cancellation of licence, the court may order that such a person shall not be eligible for a licence under the Arms Act, 1959, for a period of five years from the date of conviction.

Constitutional Provisions

India is the first country to make provisions for the protection and improvement of environment in its constitution. In the 42nd Amendment to the Constitution in 1976, provisions to this effect were incorporated in the Constitution of India with effect from 3rd January 1977. In the Directive Principles of State Policy in Chapter IV of the Constitution, Article 48-A was inserted; this enjoins the state to make an endeavour to protect and improve the environment and to safeguard the forest and wildlife of the country. Another landmark provision in respect of environment was also inserted by the same Amendment, as one of the fundamental duties of every citizen of India. This is the provision in Article 51A (g) of the Constitution. It stipulates that it shall be the duty of every citizen of India 'to protect and improve the natural environment including forests, lakes, rivers, and wildlife and to have compassion for living creatures. Without effective legislation,the resource use, pollution control, conservation, and most fields of human activity are likely to fall into chaos and conflict. It must, therefore, be made clear that there is little point in passing laws or making international agreements if there cannot be adequate enforcement. Various forms of legislation/regulation like principles, standards, guidelines, licence, declaration, convention and protocol are not firm laws, but help lawmakers.

There were provisions already existing in various enactments to tackle environmental pollution. The Indian Penal Code, The Criminal Procedure Code, The Factories Act, The Indian Forest Act, The Merchant Shipping Act, etc. have provisions for regulation and legal action for some specific environmental issues. However, with our country's emerging environmental scenario with industrialization in the post-independence era, these were found to be either inadequate or not effectively applicable to check the degradation of our environment. After the Stockholm Conference on Human Environment in June 1972, it was considered appropriate to have uniform laws all over the country. Environmental problems endanger the health and safety of our people as well as of our flora and fauna.

FOREST CONSERVATION ACT

Indian Forest Act

Although India has a Forest Act, 1927, forest-related legislation after independence came about as late as 1986 with the Forest Conservation Act, 1980. Under the Act, prior approval of central government is required before any reserved forest is declared as dereserved.

Objectives

◇ To protect and conserve forest

◇ To ensure judicial use of forest products

Under Section 3

1. Declaring that it has been decided to constitute any land as a reserve forest

2. Specifying the situation and limits of such a land and

3. Appointing an officer not holding any other forest office except that of forest settlement officer.

Under Section 4

Forest officer will perform duties of a forest settlement officer.

Powers of the State Government

Under Section 3.2, the state government is empowered to make rules for Protected Forests to regulate.

- ◇ protection and management of any portion of a forest.
- ◇ the cutting, sawing, conversion and removal of trees and timber, and removal of forest produces from such protected forests.
- ◇ the cutting of grass, and pasturing of cattle in such forests.
- ◇ hunting, shooting, fishing, setting traps, and poisoning water or snakes in such protected forests.
- ◇ granting of licenses to inhabitants living in the vicinity of protected forests to take trees, timber and forest produces for their own use.
- ◇ granting of licences to persons and the payment (if any) to be made for felling or removing trees or timber and other forest produces for trade purposes.
- ◇ inspection of forest produces for passing out from these protected forests.
- ◇ clearing and breaking-up of land for cultivation or other purposes in such protected forests.
- ◇ protection from fire of timber lying in these protected forests.

Under Section 3.5 of the Act, the State Government may prohibit

1. Pasturing of cattle
2. Cleaning of the vegetation
3. Backing up or cleaning of land for cultivation

Under Section 3.3 and the Act, an imprisonment of six months or more or a fine of Rs. 500 or more or both is the penalty for any person who violates any of the provision of this Act.

The following acts are prohibited under Section 3.

1. Clearing land in any protected forest for cultivation
2. Carrying out quarrying of any stone or burning any lime or charcoal or collecting or removing any forest produce
3. Permits cattle to damage any reserved tree.

WATER ACT

The Water (Prevention and Control of Pollution) Act, 1974, Water (Prevention and Control of Pollution) Cess Act, 1977, Air (Prevention and Control of Pollution) Act, 1981 and Environmental (Protection) Act, 1986 were enacted. Over the years, several amendments have also been made in the various existing statutes to meet the requirements of the unfolding environment issues.

Water Conservation in Agriculture

Irrigation generally makes inefficient use of water. Agricultural water used is lost due to leaks in canals, by evaporation, in barren lands where no growth occurs. The following measures can be taken to conserve water in agriculture.

- ◊ In agriculture, drip irrigation is coming into use. One of the most innovative methods in agriculture for water conservation is drip irrigation. This reduces the water needed to irrigate crops by 40 to 60%.

- ◊ Irrigation during morning or night can be done to reduce evaporation.

- ◊ Carry out better farming techniques like mining, tillage, leaving crop residues on fields, and intercropping.

- ◊ Use mulch to help retain water around the plant.

- ◊ Fix rate to use water for agriculture.

- ◊ Use crops that require less water and are drought-resistant.

- ◊ Use lined canals that reduce seepage and evaporation.

WATER CONSERVATION IN INDUSTRY

- ◊ Reuse the cooling water for irrigation and other purposes.
- ◊ Use dry cooling systems or cooling towers.
- ◊ Industries should discourage withdrawal of water wherever possible.
- ◊ Treat the wastewater and the released water can be recycled.
- ◊ Develop new equipment and processes that require less water.
- ◊ Recycling water can be used for floor washing.

SUSTAINABLE WATER CONSERVATION

- Water scarcity problem can be reduced by rain water harvesting
- Mining domestic water consumption
- Improved irrigation methods
- Recycling of wastewater
- Training water conservation.

ISSUES IMPORTANT FOR ENVIRONMENTAL LEGISLATION

1. The precautionary principle
2. The polluter-pays principle

The precautionary principle   This principle has evolved to deal with risks and uncertainties faced by environmental management. The principle implies that an ounce of prevention is worth a pound of cure—it does not prevent problems but may reduce their occurrence and helps ensure contingency plans are made.

The polluter-pays principle   The polluter pays for the damage caused by a development—this principle also implies that a polluter pays for monitoring and policing. Developing nations are seeking to have developed countries pay more for carbon dioxide and other emission controls, arguing that they polluted the global environment during the Industrial Revolution, yet enjoy the fruits of invention from that era.

BREAKING LAWS

Environmental laws are stringent in nature. They prohibit acts that bring immediate gains which, in the long run, may prove dangerous to life on this planet. Many countries have agreed to limit the emissions of greenhouse gases to the atmosphere through various sources, but when it comes to penalizing those countries exceeding their agreed limit, then one hears about possible trading from those countries that emit less. This is how countries break law.

PUBLIC AWARENESS

◇   Building on the positive links between economic growth and environment, while removing the negative links between them, requires the involvement of people in decision about how the resources will be utilized.

◇   Public participation helps the planners in better understanding of local values, knowledge and experience, helps in winning the community backing for project objectives and community help with local implementation, and can help in resolving conflicts over resources use.

◇   Individuals can involve themselves in many ways in the process of improving the environment.

Nowadays, there are many local organizations or local chapters of national/international organizations that are concerned with a variety of environmental issues/problems. People looking for a way to help can seek the organization that takes the philosophical approach most consistent with their own and that focuses on the issues that have the most meaning to them.

REVIEW QUESTIONS

1. Define sustainable development.
2. Explain the goals of sustainable development and its achievement.
3. Explain about Rio Declaration.
4. Give some reasons for energy demand.
5. Describe about the sustainable utilization of energy.
6. What is watershed management and its conservation?
7. Give detailed explanation on the factors affecting watersheds.
8. Write short notes on water conservation.
9. Explain in detail about rainwater harvesting.
10. Write short notes on rehabilitation and resettlement.
11. Define environmental ethics.
12. Explain the code of environmental ethics.
13. Discuss about the policies for climate change mitigation.
14. What is greenhouse effect?
15. List out some of the greenhouse gases.
16. Discuss about the nuclear accidents that have occurred so far.
17. Write short notes on holocaust in Bhopal.
18. What is reclamation of wastelands?
19. What are the problems associated with the development of wastelands?
20. Define consumerism.
21. What is Piguouvian tax?
22. Discuss simple ways to reduce waste.
23. Explain about Environmental Protection Act.
24. Write short notes on Air Act, Water Act, Wild Life Act and Forest Act.

25. Write short notes on population explosion and its consequences.
26. Describe the effects of population growth and its consequence on human ecology.
27. Write short notes on family welfare programme.
28. What are the environmental factors affecting human health?
29. List out some of the waterborne diseases.
30. Write short notes on the effect of air pollutants on human health.
31. What are the control measures for air pollution?
32. Give a brief description on the human rights in India.
33. What are the problems associated with human rights?
34. What is the need for value education?
35. What is HIV? Describe about the methods of its prevention.
36. What are the clinical manifestations of HIV (AIDS)?
37. What are the programmes associated with the development of women and children in India?
38. Describe about the role of IT in environment and human health.

GLOSSARY

Abiotic components Non-living components of an organism's environment, such as temperature, light, moisture, air currents, etc.

Afforestation The process of establishing a forest on land that is not a forest, or has not been a forest for a long time, by planting trees or their seeds. The term may also be applied to the legal conversion of land into the status of royal forest.

Biodiversity Biodiversity is the variation of life forms within a given ecosystem, biome, or the entire earth. Biodiversity is often used as a measure of the health of biological systems. The biodiversity found on Earth today consists of many millions of distinct biological species, which is the product of nearly 3.5 billion years of evolution.

Biogeochemical cycle In ecology and earth science, a biogeochemical cycle or nutrient cycle is a pathway by which a chemical element or molecule moves through both biotic (biosphere) and abiotic (lithosphere, atmosphere, and hydrosphere) compartments of Earth. In effect, the element is recycled, although in some cycles there may be places (called reservoirs) where the element is accumulated or held for a long period of time. Elements, chemical compounds, and other forms of matter are passed from one organism to another and from one part of the biosphere to another through the biogeochemical cycles.

Biogeographic zones Internationally recognized, large distinctive units of similar ecology, biome representation, community and species.

Biomass As a renewable energy source, it refers to living and recently dead biological material that can be used as fuel or for industrial production. In this context, biomass refers to plant matter grown to generate electricity or produce for example trash such as dead trees and branches, yard clippings and wood chips biofuel, and it also includes plant or animal matter used for production of fibres, chemicals or heat. Biomass may also include biodegradable wastes that can be burnt as fuel. It excludes organic material which has been transformed by geological processes into substances such as coal or petroleum.

Biomining A new approach to the extraction of desired minerals from ores being explored by the mining industry in the past few years. Microorganisms are used to leach out the minerals, rather than the traditional methods of extreme heat or toxic chemicals, which have a deleterious effect on the environment.

Biosphere   It is the global sum of all ecosystems. From the broadest biophysiological point of view, the biosphere is the global ecological system integrating all living beings and their relationships, including their interaction with the elements of the lithosphere, hydrosphere, and atmosphere. This biosphere is postulated to have evolved, beginning through a process of biogenesis or biopoesis, at least some 3.5 billion years ago.

Biotic components   The living things that shape an ecosystem. They are in entirety, any living component that affects another organism. Such things include animals which consume the organism in question, and the living food that the organism consumes.

cDNA library   A collection of cloned cDNA (complementary DNA) fragments inserted into a collection of host cells, which together constitute some portion of the transcriptome of the organism.

Conservation movement   A political, social and, to some extent, scientific movement that seeks to protect natural resources including plant and animal species as well as their habitat for the future. Also known as nature conservation.

DDT   One of the best known synthetic pesticides (**D**ichloro-**D**iphenyl-**T**richloroethane).

Deforestation   The logging or burning of trees in forested areas. There are several reasons for doing so: trees or derived charcoal can be sold as a commodity and are used by humans while cleared land is used as pasture, plantations of commodities and human settlement. The removal of trees without sufficient reforestation has resulted in damage to habitat, biodiversity loss and aridity. Also deforested regions often degrade into wasteland.

Desertification   The degradation of land in arid and dry sub-humid areas, resulting primarily from natural activities and influenced by climatic variations. It is also a failure of the ecological succession process.

Drought   An extended period of months or years when a region notes a deficiency in its water supply. Generally, this occurs when a region receives consistently below-average precipitation.

Earth Summit   United Nations Conference on Environment and Development (UNCED) at Rio de Janeiro, 3–14 June, 1992.

Ecosystem   A natural unit consisting of all plants, animals and micro-organisms (biotic factors) in an area functioning together with all of the non-living physical (abiotic) factors of the environment. An ecosystem is a completely independent unit of interdependent organisms which share the same habitat. Ecosystems usually form a number of food webs which show the interdependence of the organisms within the ecosystem.

Endangered species   The population of organisms which is at a risk of becoming extinct because it is either few in number, or threatened by changing environmental or predation parameters. An endangered species is usually a taxonomic species, but may be another evolutionarily significant unit.

Endemism   The ecological state of being unique to a particular geographic location, such as a specific island, habitat type, nation, or other defined zone. To be endemic to a place or area means that it is found only in that part of the world and nowhere else.

Environment   A system is the part of the universe that is being studied, while the environment is the remainder of the universe that lies outside the boundaries of the system. It is also known as the surroundings.

Environmental remediation   The removal of pollution or contaminants from environmental media such as soil, ground water, sediment, or surface water for the general protection of human health and the environment or from a brownfield site intended for re-development.

Erosion   It is the removal of solids (sediment, soil, rock and other particles) in the natural environment. It usually occurs due to transport by wind, water, or ice; by down-slope creep of soil and other material under the force of gravity; or by living organisms, such as burrowing animals, in the case of bioerosion.

Genetic engineering   A recombinant DNA technology, genetic modification/ manipulation (GM) and gene splicing are terms that apply to the direct manipulation of an organism's genes. Genetic engineering is different from traditional breeding, where the organism's genes are manipulated indirectly. Genetic engineering uses the techniques of molecular cloning and transformation to alter the structure and characteristics of genes directly.

Geothermal power   The power extracted from heat stored in the earth. This geothermal energy originates from the original formation of the planet, from radioactive decay of minerals, and from solar energy absorbed at the surface. It has been used for space heating and bathing since ancient roman times, but is now better known for generating electricity.

Gobar gas   As the gobar gas production is an anaerobic process, it is carried out in an air tight, closed cylindrical concrete tank called a digester. The tank has a concrete inlet basin on one side for feeding fresh cattle dung (gobar). There is a concrete outlet on the outer side for removing the digested sludge. The top of the tank serves as the gas tank. It has an outlet pipe for the gobar gas.

Green Revolution   The term "Green Revolution" is a general one that is applied to successful agricultural experiments in many Third World countries. It is not specific to India. But it was most successful in India. There were three basic elements in the method of the Green Revolution: (1) Continued expansion of farming areas; (2) Double-cropping of existing farmland; (3) Using seeds with improved genetics.

Irrigation   An artificial application of water to the soil usually for assisting in growing crops. In crop production it is mainly used in dry areas and in periods of rainfall shortfalls, but also to protect plants against frost. Additionally irrigation helps to suppress weed growing in rice fields.

Landslide   A geological phenomenon which includes a wide range of ground movement, such as rock falls, deep failure of slopes and shallow debris flows, which can occur in offshore, coastal and onshore environments.

Livestock   The term used to refer (singularly or plurally) to a domesticated animal intentionally reared in an agricultural setting to produce things such as food or fibre, or for its labour.

Magma   Molten rock that is found beneath the surface of the Earth, and may also exist on other terrestrial planets. Besides molten rock, magma may also contain suspended crystals and gas bubbles. Magma often collects in a magma chamber inside a volcano. Magma is capable of intrusion into adjacent rocks, extrusion onto the surface as lava, and explosive ejection as tephra to form pyroclastic rock.

Mechanization   The process of providing human operators with machinery to assist them with the physical requirements of work. It can also refer to the use of machines to replace manual labour or animals. A step beyond mechanization is automation. The use of hand-powered tools is not an example of mechanization.

Metabolism   It is the set of chemical reactions that occur in living organisms in order to maintain life. These processes allow organisms to grow and reproduce, maintain their structures, and respond to their environments. Metabolism is usually divided into two categories. Catabolism breaks down organic matter, for example to harvest energy in cellular respiration. Anabolism, on the other hand, uses energy to construct components of cells such as proteins and nucleic acids.

Mineral   A naturally occurring solid formed through geological processes, and has a characteristic chemical composition, a highly ordered atomic structure, and specific physical properties.

Mining   The extraction of valuable minerals or other geological materials from the earth, usually from an ore body, vein or (coal) seam. Materials recovered by mining include base metals, precious metals, iron, uranium, coal, diamonds, limestone, oil shale, rock salt and potash. Any material that cannot be grown through agricultural processes, or created artificially in a laboratory or factory, is usually mined. Mining in a wider sense comprises extraction of any non-renewable resource (e.g. petroleum, natural gas, or even water).

Nutrient   The chemical that an organism needs to live and grow or a substance used in an organism's metabolism which must be taken in from its environment. Organic nutrients include carbohydrates, fats, proteins (or their building blocks, amino acids), and vitamins. Inorganic chemical compounds such as dietary mineral, water and oxygen may also be considered nutrients. A nutrient is essential to an organism if it cannot be synthesized by the organism in sufficient quantities and must be obtained from an external source. Nutrients needed in relatively large quantities are called macronutrients and those needed in relatively small quantities are called micronutrients.

Overgrazing A phenomenon that occurs when plants are exposed to intensive grazing for extended periods of time, or without sufficient recovery periods. It can be caused by either livestock in poorly managed agricultural applications, or by overpopulations of native or non-native wild animals.

Polychlorinated dibenzodioxins A group of polyhalogenated compounds which are significant because they act as environmental pollutants. They are commonly referred to as dioxins for simplicity in scientific publications because every PCDD molecule contains a dioxin skeletal structure. Typically, the *p*-dioxin skeleton is at the core of a PCDD molecule, giving the molecule a dibenzo-*p*-dioxin ring system. Members of the PCDD family have been shown to bioaccumulate in humans and wildlife due to their lipophilic properties, and are known teratogens, mutagens, and suspected human carcinogens. They are organic compounds. Dioxins occur as by-products

in the manufacture of organochlorides, in the incineration of chlorine-containing substances such as PVC (polyvinyl chloride), in the bleaching of paper, and from natural sources such as volcanoes and forest fires.

Resource Any physical or virtual entity of limited availability, or anything used to help one earn a living. In most cases, commercial or even ethical factors require resource allocation through resource management.

Sustainable development A pattern of resource use that aims to meet human needs while preserving the environment so that these needs can be met not only in the present, but also for future generations to come. The term was used by the Brundtland Commission which coined what has become the most often-quoted definition of sustainable development as development that "meets the needs of the present without compromising the ability of future generations to meet their own needs.

Topography The study of earth's surface shape and features or those of planets, moons, and asteroids. It is also the description of such surface shapes and features (especially their depiction in maps).

BIBLIOGRAPHY

Agarwal, A. and Narain, S. (2000). "Water harvesting: community–led natural resource management." *LEISA. ILEIA Newsletter.* 16(1): 11–13.

Aguilera-Klink, F., Perez-Moriana, E. and Sanchez-Garcia, J. (2000). "The social construction of scarcity. The case of water in Tenerife (Canary Island)." *Ecological Economics.* 34: 233–245.

Ahluwalia, M. (1997). "Representing communities: the case of a community-based watershed management project in Rajasthan, India." *IDS Bulletin.* 28(4): 23–35.

Alnutt, T., Wikramanayake, E., Dinerstein, E., Loucks, C., Jackson, R., Hunter, D. and Carpenter, C. (2002). "Composition of the alpine Himalayan protected areas network and its contribution to biodiversity conservation." In: Wikramanayake, E.D., Dinerstein, E., Loucks, C., Olson, D., Morrison, J., Lamoreux, J., McKnight, M. and Hedao, P. (eds.). *Terrestrial Ecoregions of the Indo-Pacific: A Conservation Assessment.* Island Press, Washington, D.C. pp. 131–135.

Anderson, T.W. and Runge, C.F. (1994). "Common Property and Collective Actions: Lessons from Cooperative Watersheds Management in Haiti".

Arya, S.L. and Samra, J.S. (1996). "Determinants of People's Participation in Watershed Development and Management: An Exploratory Case Study in Shivalik Foothill Villages in Haryana." In: Iyer, K.G. (ed.). *Sustainable Development: Ecological and Sociocultural Dimensions.* Vikas Publishing House, New Delhi.

Ashton, P.S. and Gunatilleke, C.V.S. (1987). "New light on the plant geography of Ceylon I: Historical plant geography." *Journal of Biogeography.* 14: 249–285.

Bakker, M., Barker, R., Meinzen-Dick, R.S. and Konransen, F. (eds.). (1999). "Multiple uses of water in irrigated areas: A case study from Sri Lanka." *SWIM Report 8*, International Water Management Institute, Colombo, Sri Lanka.

Bardhan, P. (2000). "Irrigation and cooperation: and empirical analysis of 48 irrigation communities in South India." *Economic Development and Cultural Change*. 48-49(1-4): 847–66.

Benvenuti, D. (1988). "Community participation in soil and water conservation." In: Moldenhauer, W.C. and Hudson, N.W. (eds.). *Conservation Farming on Steep Lands, Soil and Water Conservation Society*. IA.

Biju, S.D. and Bossuyt, F. (2003). "New frog family from India reveals an ancient biogeographical link with the Seychelles." *Nature*. 425:711–714.

Biodiversity Action Plan. (2002). Biodiversity Action Plan for Bhutan 2002. Prepared and published by the Ministry of Agriculture, Royal Government of Bhutan.

BirdLife International. (2003). *Saving Asia's Threatened Birds: A Guide for Government and Civil Society*. Cambridge.

Botanical Survey of India. (1983). *Flora and Vegetation of India-An Outline*. Botanical Survey of India, Howrah. p. 24.

Burroughs, R. (1999). "When stakeholders choose: process, knowledge, and motivation in water quality decisions." *Society and Natural Resources*. 12(8): 797–809.

Bustard, H.R. (1982). "Crocodile breeding project." In: *Wildlife in India*. Saharia, V.B. (ed.). Natraj Publishers, Dehradun. pp. 147–163.

Champion, H.G. (1936). "A preliminary survey of the forest types of India and Burma." *Indian Forest Record (New Series)*. 1: 1–286.

Champion, H.G. and Seth, S.K. (1968). *"A Revised Survey of the Forest Types of India*. Govt of India Press, Delhi. p. 404.

Chopra, K., Kadekodi, G.K. and Murty, M.N. (1988). "Sukhomajri and Dhamala watersh eds in Haryana: a participatory approach to management." Institute of Economic Growth, Delhi, India. (Working Paper).

Collins, N.M., Sayer, J. and Whitmore, T.C. (eds.). (1991). *The Conservation Atlas of Tropical Forests: Asia and the Pacific*. IUCN, Gland, Switzerland and Cambridge, UK. pp. 256.

Conservation International. (2000). *Biodiversity Hotspots* (a map).

Conti, E., Rutschmann, F. and Eriksson, T. (2003). Testing the out-of-India origin of Crypteroniaceae with molecular dating methods [Abstract].

In: Proceedings of the Annual Meeting of the Association for Tropical Biology and Conservation. University of Aberdeen, Aberdeen, U.K. p. 37.

Dahanukar, N., Raut, R. and Bhat, A. (2004). "Distribution, endemism and threat status of freshwater fishes in the Western Ghats of India." *Journal of Biogeography*. 31: 123–136.

Dani, A.A. and Campbell, J.G. (1986). "Sustaining upland resources: people's participation in watershed management." Occasional Paper No. 3, International Centre for Integrated Mountain Development (ICOMOD), Kathmandu, Nepal.

Daniels, R.J.R. (2001). "Endemic fishes of the Western Ghats and the satpura hypothesis." *Current Science*. 81:241–244.

De Alwis, Lyn. (1983). "A Nation Rises to the Challenge." *National Geographic*. 274–278.

Dhar, S.K. (1994). "Rehabilitation of degraded tropical forest watersheds with people's participation." Joint Forest Management Series No. 16, Haryana Forest Department and Tata Energy Research Institute, New Delhi, India.

Environmental Encyclopedia. (1994). Gale Research International Limited, Detroit.

FAO/UNEP (1981). Tropical forest resources assessment project. Technical report No. 3. FAO, Rome.

Farington, J., Turton, C. and James, A.J. (eds.). (1999). *Participatory Watershed Development: Challenges for the Twenty-first Century*." Oxford University Press, New Delhi.

Farrington, J. and Lobo, C. (1997). "Scaling up participatory watershed development in India: lessons from the Indo-German watershed development programme." Natural Resource Perspectives No. 17, Overseas Development Institute, London.

Gadgil, M. (1985). "Social Restraints on Resource Utilization: The Indian Experience." In: McNeely, J.A. and Pitt, D. (eds.). *Culture and Conservation: The Human Dimension in Environmental Planning*. Dublin Croom, Helm. pp. 135–155.

Gadgil, Madhav (1989). "The Indian heritage of a conservation ethic." In: *Conservation of the Indian Heritage*. Allchin, B., Allchin, F.R. and Thapar, B.K. (eds.). Cosmo Publishers, New Delhi. pp. 13–21.

Galvez, J.B. and Macatumbas, D.B. (1993). "Community-Based Resource Management: Perspectives, Experiences and Policy Issues Related to Irrigation." In: Fellizar, F.P. (eds.). Community-Based Resource Management: Perspectives, Experiences and Policy Issues, Report No. 6, Environment and Resource Management Project (ERMP), Laguna, Philippines.

Ganeshaiah, K.N. (2003). *Sasya Sahyadri: Distribution, Taxonomy and Diversity of Plants of Western Ghats*. University of Agricultural Sciences, Bangalore, India.

Ghosh, A.K. (1996). Faunal Diversity. In: Gujral, G.S. and Sharma, V. (eds.). *Changing Perspectives of Biodiversity Status in the Himalaya*. The British Council, New Delhi.

Gimaret-Carpentier, C., Dray, S. and Pascal, J.P. (2003). "Broad-scale biodiversity pattern of the endemic tree flora of the Western Ghats (India) using canonical correlation analysis of herbarium records." *Ecography.* 26: 429–444.

Government of India (1985). Research and Reference Division Ministry of Information and Broadcasting.

Grewal, S.S. (1996). "Community participation in watershed development." In: Iyer, K.G. (ed). *Sustainable Development: Ecological and Sociocultural Dimensions*. Vikas Publishing House, New Delhi.

Grimmet, R., Inskipp, C. and Inskipp, T. (1999). *Birds of India, Pakistan, Nepal, Bangladesh, Bhutan, Sri Lanka, and the Maldives. Princeton Field Guides*. Princeton University Press, Princeton.

Groombridge, B. (1983). Comments on the rain forests of southwest India and their herpetofauna. Paper prepared for the Centenary Seminar of the Bombay Natural History Society, 6–10 December, 1983. p.18. Revised, January 1984.

Groombridge, B. (ed.). (1993). *The 1994 IUCN Red List of Threatened Animals*. IUCN, Gland, Switzerland and Cambridge, UK. p. 286.

Gupta, R.K. (1994). "Arcto-alpine and boreal elements in the high altitude flora of the northwest Himalaya." In: Pangtey, Y.P.S. and Rawal, R.S. (eds.). *High Altitudes of the Himalaya. (Biogeography, Ecology and Conservation)*. Gyanodaya Prakashan, Delhi. pp. 12–32.

Haribal, M. (1992). *Butterflies of Sikkim Himalaya and their Natural History*. Nature Conservation Foundation, Gangtok, Sikkim.

Hinchcliffe, H. (ed.). (1999). *Fertile Ground: Impacts of Participatory Watershed Management*. IT Publications, London.

IBWL (1972). Project Tiger. A Planning Proposal for Preservation of Tiger *(Panthera tigris tigris Linn.)* in India. Indian Board for Wildlife, Government of India, New Delhi. p.114.

ICBP (1992). Putting Biodiversity on the Map: Priority Areas for Global Conservation. International Council for Bird Preservation, Cambridge, UK. p.90.

IUCN (1986). *Review of the Protected Areas System in the Indo-Malayan Realm*. IUCN, Gland, Switzerland and Cambridge, U.K. p. 461.

IUCN (1987). Centres of Plant Diversity: A Guide and Strategy for their Conservation (An outline of a book being prepared by the Joint IUCN-WWF Plants Conservation Programme and IUCN Threatened Plants Unit).

IUCN. (1990). *IUCN Directory of South Asian Protected Areas*. IUCN, Cambridge.

Jayaraman, T.K. (1980). "People's participation in the implementation of watershed management projects: An empirical study from Gujarat." *Indian Journal of Public Administration*. 26(4): 1009–1016.

Jha, C.S., Dutt, C.B.S. and Bawa, K.S. (2000). "Deforestation and land use changes in Western Ghats." India. *Current Science*. 79:231–238.

Jonish, J. (1993). "Institution and Collective action: self–governance in irrigation." *Review of Social Economy*. 51(1): 116(3), Spring 1993.

Kar, C.S. and Bhaskar, S. (1981). "Status of sea turtles in the Eastern Indian Ocean. In: *Biology and Conservation of Sea Turtles*. Bjorndal, K. (ed.). Proc. World Conf. Sea Turtle Cons., Smithsonian Institute Press, Washington. pp. 373–383.

Kerr, J. (2001). "Watershed project performance in India: conservation, productivity, and equity." *American Journal of Agricultural Economics*. 83(5): 1223–1230.

Kikuchi, M., Weligamage, P., Barker, R., Samad, M., Kon, H. and Somaratne, H.M. (2003). "Agro-well and Pump Diffusion in the Dry Zone of Sri Lanka." Report No.66, IWMI Research, International Water Management Institute, Colombo, Sri Lanka.

Kunte, K., Joglekar, A., Ghate, U. and Pramod, P. (1999). "Patterns of butterfly, birds and tree diversity in the Western Ghats." *Current Science*. 77(4): 577–586.

Lal, J.B. (1989). *India's Forests: Myth and Reality.* Natraj Publishers, New Delhi, India.

Lobo, C. (1995). "Rain decided to help us: Participatory watershed management in the State of Maharashtra, India." Tata Energy Research Institute, New Delhi.

MacKinnon, J. and MacKinnon, K. (1986). *Review of the Protected Areas System in the Indo-Malayan Realm.* International Union for the Conservation of Nature and Natural Resources, Gland, Switzerland and Cambridge, U.K. p. 284.

Meegaskumbura, M., Bossuyt, F., Pethiyagoda, R., Manamendra-Arachchi, K., Milinkovitch, M.C. and Schneider, C.J. (2002). "Sri Lanka: An amphibian hotspot." *Science.* 298: 379.

Meinzen-Dick, R.S., Raju, K.V. and Gulati, A. (2000). "What affects organization and collective action for managing resources? Evidence form canal irrigation systems in India." EPTD Discussion Paper 61, International Food Policy Research Institute, Washington, D.C.

Mittal, S.P. (1996). "Land and water management for sustainable production with people's participation." In: Iyer, K.G. (ed.). *Sustainable Development: Ecological and Sociocultural Dimensions,* Vikas Publishing House, New Delhi.

Mittermeier and Russell, *et al.* (1999). *Hotspots.* Agrupacion Sierra Madre.

Mittermeier, R.A., Carr, J.L. and Swingland, I.R. *et al.* (1992). "Conservation of amphibians and reptiles." In: Adler, K. (ed.). *Herpetology: Current Research on the Biology of Amphibians and Reptiles.* Society for the Study of Amphibians and Reptiles, Ithaca.

Mollinga, P.P. (2001). "Water and politics: levels, rational choice and South Indian canal irrigation." Futures, Oct-Nov 2001, 733.

Myers, Norman. (1988). "Threatened Biotas: 'Hot Spots' in Tropical Forests." *The Environmentalist.* Vol. 8. 3: 187–208.

Myers, Norman. (1990). "The Biodiversity Challenge: Expanded Hot-Spots Analysis." *The Environmentalist.* Vol. 10. 4: 243–256.

Naggs, F. and Raheem, D. (2000). Land Snail Diversity in Sri Lanka. CD ROM. Natural History London: Museum.

Naggs, F. and Raheem, D. The Darwin Initiative Project on Sri Lankan land snails: Patterns of diversity in Sri Lankan forests" In: *Molluscan*

Conservation and Biodiversity. Killeen, I. and Seddon, M. (eds.). Conchological Society, London. In press.

Narayan, D. (1997). "Focus on people's participation: evidence from 121 rural water projects." In: Clague, C. (ed.). Institutions and Economic Development: Growth and Governance in Less-Developed and Post-Socialist Countries, The Johns Hopkins Studies in Development, Johns Hopkins University Press, Baltimore.

Navalawala, B.N. (1995). "Participatory irrigation management in India: status, issues and approach." In: Geijer, J. (ed.). Irrigation Management Transfer in Asia: Papers from the Expert Consultation on Irrigation Management Transfer in Asia, Food and Agriculture Organization of the United Nations, International Irrigation Management Institute, Bangkok, Thailand, (RAP Publication 1995: 31).

Nayar, M.P. and Sastry, A.R.K. (eds.). (1987). *Red Data Book of Indian Plants,* Vol. 1. Botanical Survey of India, Calcutta. p. 367.

Ng, P.K.L. and Tay, F.W.M. (2001). The freshwater crabs of Sri Lanka (Decapoda: Brachyura: Parathelphusidae). *Zeylanica.* 6(2): 113–199.

Olson, M. David. and Eric Dinerstein. (1998). The Global 200: A Representation Approach to Conserving the Earth's Distinctive Ecoregions. World Wildlife Fund.

Panwar, H.S. (1982). "Project Tiger." In: *Wildlife in India.* Saharia, V.B. (ed.). Natraj Publications, Dehradun. pp. 130–137.

Parthasarathy, R. and Iyengar, S. (1998). "Participatory Water Resources Development in Western India." In: Mosse, D., Farrington, J. and Re weds, A. Development as Process:Concepts and Methods for Working with Complexity, ODI Development Studies, No. 2, Routledge and Overseas Development Institute, New York, (Routledge Research).

Pascal, J.P. (1988). Wet Evergreen Forests of the Western Ghats of India: Ecology, Structure, Floristic Composition and Succession. French Institute, Pondicherry, India.

Pepper, L. Ian, Gerba, P. Charles and Brusseau, L. Mark. (1996). *Pollution Science.* Academic Press, New York. .

Pethiyagoda, R. (1991). *Freshwater Fishes of Sri Lanka.* Wildlife Heritage Trust, Colombo.

Pethiyagoda, R. (1994). "Threats to the indigenous freshwater fishes of Sri Lanka and remarks on their conservation." *Hydrobiologia.* 285:189–201.

Pethiyagoda, R. and Manamendra-Arachchi, K. (1998). "Evaluating Sri Lanka's amphibian diversity." *Occasional Papers of the Wildlife Heritage Trust of Sri Lanka.* 2: 1–12.

Pillai, V.N.K. (1982). "Status of wildlife conservation in states and union territories." In: Saharia, V.B. (ed.). *Wildlife in India.* Natraj Publishers, Dehradun. pp. 74–91.

Plage, Dieter and Plage, Mary. (1983). "Legacy of lively treasures." *National Geographic.* 256–273.

Potkanski, T. and Adams, W.M. (1998). "Water scarcity, property regimes and irrigation management in Sonjo, Tanzania." *Journal of Development Studies.* 34(4) : 86(31).

Price, H.D. (1995). "The cultural effects of conveyance loss in gravity-fed irrigation systems." *Ethnology.* 34 (4): 273–91.

Putman, J. John, (1976). "India struggles to save her wildlife." *National Geographic.* 298–343.

Ramesh, B.R. and Pascal, J.P. (1991). Distribution of Endemic, arbsorescent Evergreen Species in the Western Ghats. Proceedings of Symposium on Rare and Endangered Plants of the Western Ghats Kerala Forest Department. Trivandrum.

Rhoades, R.E. (1998). "Participatory Watershed Research and Management: Where the Shadow Falls." Sustainable Agriculture Programme, International Institute for Environment and Development, London.

Rodgers, W.A. and Panwar, H.S. (1988). *Planning a Wildlife Protected Area Network in India.* 2 vols. Project FO: IND/82/003. FAO, Dehradun. 339, p. 267.

Rodgers, W.A., Panwar, H.S. and Mathur. V.B. (2002). Wildlife Protected Area Network in India: A review. Wildlife Institute of India, Dehradun.

Roelants, K., Jiang, J. and Bossuyt, F. Endemic ranid (Amphibia: Anura) genera in southern mountain ranges of the Indian subcontinent represent ancient frog lineages: Evidence from molecular data. Molecular Phylogenetics and Evolution. In press.

Sakurai, T. and Palanisami, K. (2001). "Tank irrigation management as a local common property: the case of Tamil Nadu, India." *Agricultural Economics.* 25(2-3): 273–283.

Salm, R.V. (1981). Coastal resources in Sri Lanka, India and Pakistan: description.

Samra, J.S. and Dhyani, B.L. (1998). "Elements of Participatory Watershed Management in India. In: National Workshop on Watershed Approach for Managing Degraded Lands in India Challenges for 21st century, Sponsored by Government of India, Ministry of Rural Areas and Employment Department of Wastelands Development, New Delhi.

Scidmore, Eliza Ruhamah. (1907). "The bathing and burning ghats at Benares." *National Geographic*. 118–128.

Scott, D.A. (1989). *A Directory of Asian Wetlands*. IUCN, Gland, Switzerland, and Cambridge, UK.

Shah, P. (1994). "Participatory watershed management in India: the experience of the Aga Khan Rural Support Programme." In: *Beyond Farmer First*. Scoones, I. and Thompson, J. (eds.). Intermediate Technologies Publications, London.

Siddall, M., Rohling, E.J. and Almogi-Labin, A. *et al.* (2003). "Sea-level fluctuations during the last glacial cycle." *Nature*. 423:853–858.

Singh, K. (1991). "Determinants of people's participation in watershed development and management: An exploratory case study." *Indian Journal of Agricultural Economics*. 46(3): 278–286.

Singh, K. (1992). "Managing Common Pool Irrigation Tanks: A Case Study in Andhra Pradesh and West Bengal." Case Study No. 9, Institute of Rural Management, Anand, India.

Sukumar, R. (2003). *"The Living Elephants: Evolutionary Ecology, Behaviour, and Conservation*. Oxford University Press, New York.

Tewari, D.N. (1995). *Western Ghats Ecosystem*. International Book Distributors.

UNDP. (1998). Ecoregional Co-Operation for Biodiversity Conservation in the Himalayas. Proceedings of a regional meeting organized by UNDP in co-operation with WWF and ICIMOD.

UNEP/IUCN (1988). *Directory of Coral reefs of International Importance. Vol. 2. Indian Ocean, Red Sea and Gulf*. UNEP Regional Seas Directories and Bibliographies. IUCN, Gland, Switzerland, Cambridge, UK/UNEP, Nairobi, Kenya. 389 pp, 36 maps.

Uphoff, N. (1985). "People's participation in water management: Gal Oya, Sri Lanka." In: Garcia-Zamov, J. (eds.). *Public Participation in Development Planning and Management*, Westview, Boulder, CO.

Wade, R. (1995). "The Ecological Basis of Irrigation Institutions: East and South Asia." *World Development.* 23(12): 2041–2049.

Walter, E.C. and Ahmed, B. (1979). "Village, Technology, and Bureaucracy: Patterns of Irrigation Organization in Comilla District, Bangladesh." *The Journal of Developing Areas.* 13: 431–440.

Ward, C. Geoffrey. (1992). "India's wildlife dilemma." *National Geographic.* 2–29.

Wilson, E.O. (1992). *The Diversity of Life.* W.W. Norton and Co.

WWF and ICIMOD. (2001). Ecoregion-based Conservation in the Eastern Himalaya. Identifying Important Areas for Biodiversity Conservation. WWF Nepal Program, Kathmandu.

WWF Nepal. (2004). WWF Nepal Annual Report. Nepal, Kathmandu.

Web References

www.tiscale.co.uk/reference/encyclopedia/hutchinson/m0023675.html

www.nrel.gov/rredc/

www.sandeeonline.org

www.library.thinkquest.org/26026/Science.

www.nature.nps.gov/.

www.csm.jmu.edu/biology/humeyca/reliable resources/typesof resources.htm

www.ses.sk.cu/vol/HTT/RR/minerals.htm

www.conservation.org/xp/CIWEB/strategies/hotspots/hotspots.xml

www.ces.lise.ernet.in/hpg/cesmg/pew/wgbn.html

www.kudre mukh.org

www.worldwildlife.org/bsp/

www.wwf.org

www.usaid.org

www.wri.org

www.nbaindia.org

www.ces.iisc.ernet.in/hpg/cesmg/indiabio.html

www.unep.org